PYTHON MACHINE LEARNING: A BEGINNER'S GUIDE TO SCIKIT-LEARN

A Hands-On Approach

RAJENDER KUMAR

Trademarks

Screenshots

Website References

All website references were current at the date of publication.

For more information, contact: support@JambaAcademy.com.

Published by:
Jamba Academy

Printed in the United States of America

First Printing Edition, 2023

Found Typos & Broken Link

We apologize in advance for any typos or broken link that you may find in this book. We take pride in the quality of our content and strive to provide accurate and useful information to our readers. Please let us know where you found the typos and broken links (if any) so that we can fix them as soon as possible. Again, thank you very much in advance for bringing this to our attention and for your patience.

If you find any typos or broken links in this book, please feel free to email us.

support@JambaAcademy.com

Support

We would love to hear your thoughts and feedback! Could you please take a moment to write a review or share your thoughts on the book? Your feedback helps other readers discover the books and helps authors to improve their work. Thank you for your time and for sharing your thoughts with us!

If there is anything you want to discuss or you have a question about any topic of the book, you can always reach out to us, and we will try to help as much as we can.

support@JambaAcademy.com

To all the readers who have a passion for programming and technology, and who are constantly seeking to learn and grow in their field.

This book is dedicated to you and to your pursuit of knowledge and excellence.

DISCLAIMER

This book is intended for educational purposes only and is not a substitute for professional advice. The information provided in this book is accurate to the best of the author's knowledge, but the author and publisher cannot be held responsible for any errors or omissions. The author and publisher shall have neither liability nor responsibility to any person or entity with respect to any loss or damage caused or alleged to be caused directly or indirectly by the information contained in this book. The examples and case studies used in this book are for illustrative purposes only and are not intended to serve as a guarantee of success. Your results may vary depending on your specific circumstances. It is your responsibility to conduct your own research and seek the advice of a professional before making any decisions based on the information provided in this book.

ACKNOWLEDGMENTS

I would like to express my heartfelt gratitude to my colleagues, who provided valuable feedback and contributed to the development of the ideas presented in this book. In particular, I would like to thank Tanwir Khan for his helpful suggestions and support.

I am also grateful to the editorial and production team at Jamba Academy for their efforts in bringing this book to fruition. Their professionalism and expertise were greatly appreciated.

I also want to thank my family and friends for their love and support during the writing process. Their encouragement and understanding meant the world to me.

Finally, I would like to acknowledge the many experts and thought leaders in the field of data science whose works have inspired and informed my own. This book is the culmination of my own experiences and learning, and I am grateful to the wider community for the knowledge and insights that have shaped my thinking.

This book is a product of many people's hard work and dedication, and I am grateful to all of those who played a role in its creation.

HOW TO USE THIS BOOK?

"How to Use This Book" is a guide for readers to effectively navigate and make the most out of the content provided in this book. Here are a few tips on how to use the book to its fullest potential:

I. **Begin with the introduction**: Start by reading the introduction to gain an understanding of the overall purpose and structure of the book. This will help you to better understand the context and flow of the information provided.

II. **Follow the chapter sequence**: The chapters are organized in a logical sequence, building on one another to provide a comprehensive understanding of the topic. It is recommended to read the chapters in order to fully grasp the concepts presented.

III. **Utilize the examples and exercises**: The book includes examples, exercises, and case studies to help readers better understand and apply the concepts discussed. Make sure to work through them as they appear in the book.

IV. **Apply the information**: The best way to truly understand and retain the information presented in the book is to apply it in real-world scenarios. Try to use the concepts discussed in your own work or personal projects to gain hands-on experience and solidify your understanding.

V. **Review the summary and questions at the end of each chapter**: The summary and questions provided at the end of each chapter are designed to help you review and test your understanding of the material. Make sure to review them before moving on to the next chapter.

VI. **Seek help if needed**: If you find yourself struggling to understand a concept or need additional assistance, don't hesitate to reach out to others for help. Join online communities, attend meetups, or seek out a mentor to help you overcome any obstacles you may encounter.

VII. **Reference the additional resources**: The book includes various resources such as websites, books, and online courses to provide additional information and support for readers. Use these resources to supplement your learning and stay up-to-date with the latest developments in the field.

By following these tips, you will be able to use this book to its full potential and gain a comprehensive understanding of machine learning and its applications in the real world.

CONVENTIONS USED IN THIS BOOK

When learning a new programming language or tool, it can be overwhelming to understand the syntax and conventions used. In this book, we follow certain conventions to make it easier for the reader to follow along and understand the concepts being presented.

Italics

Throughout the book, we use italics to indicate a command used to install a library or package. For example, when we introduce the Keras library, we will italicize the command used to install it:

!pip install keras

Bold

We use bold text to indicate important terminology or concepts. For example, when introducing the concept of **neural networks**, we would bold this term in the text.

Handwriting Symbol

At times, we may use a handwriting symbol to indicate an important note or suggestion. For example, we may use the following symbol to indicate that a certain code snippet should be saved to a file for later use:

This is an important note or information.

Code Examples

All code examples are given inside a bordered box with coloring based on the Notepad++ Python format. For example:

```
import tensorflow as tf
from tensorflow import keras
from tensorflow.keras import layers
import numpy as np
import matplotlib.pyplot as plt

# Load the dataset
(x_train, y_train), (x_test, y_test) =
keras.datasets.cifar10.load_data()

# Keep only cat and dog images and their labels
train_mask = np.any(y_train == [3, 5], axis=1)
test_mask = np.any(y_test == [3, 5], axis=1)
x_train, y_train = x_train[train_mask], y_train[train_mask]
x_test, y_test = x_test[test_mask], y_test[test_mask]
```

Output and Explanation

Below each code example, we provide both the output of the code as well as an explanation of what the code is doing. This will help readers understand the concepts being presented and how to apply them in their own code.

Overall, by following these conventions, we hope to make it easier for readers to follow along and learn the concepts presented in this book.

GET CODE EXAMPLES ONLINE

The book "Python Machine Learning: A Beginner's Guide to Scikit-Learn" is a comprehensive guide for machine learning and deep learning concepts using Python. It covers various machine learning algorithms and deep learning architectures along with hands-on examples to get a better understanding of the concepts.

To make it even more convenient for readers, we are offering all the code discussed in the book as Jupyter notebooks on the link:

https://github.com/JambaAcademy/Python-Machine-Learning-A-Beginners-Guide-to-Scikit-Learn-Book-Code

This will allow readers to access and use the code examples easily.

Jupyter notebooks provide an interactive computing environment that enables users to write and run code, as well as create visualizations and documentation in a single document. This makes it a perfect tool for learning and experimenting with machine learning and deep learning concepts.

The code provided on the Github repository can be downloaded and used freely by readers. The notebooks are organized according to the chapters in the book, making it easier for readers to find the relevant code for each concept.

We believe that this initiative will help readers to gain a better understanding of machine learning and deep learning concepts by providing them with practical examples that they can run and experiment with.

ABOUT THE AUTHOR

Meet Rajender Kumar, an experienced data professional with over 11 years of experience in the field. Rajender Kumar has a diverse background in data science, machine learning, analysis, and data integration. With a passion for data-driven businesses, Rajender has dedicated his career to understanding and solving complex data challenges.

Throughout his career, Rajender has worked with a wide variety of clients and industries, including finance, healthcare, retail, and more. This diverse experience has given him a unique perspective on data and the ability to approach problems with a holistic mindset.

Rajender's interests go beyond just the technical aspects of his work. He is also deeply interested in the ethical and philosophical implications of artificial intelligence, and how we can use technology to benefit society in a responsible and sustainable way. This interest in the broader impacts of technology has led Rajender to explore topics such as spirituality and mindfulness, as he believes that a holistic approach to problem-solving is crucial in the rapidly evolving world of data and AI. He is a firm believer in the power of data to drive business success and is excited to share his expertise with others through this book.

With a wealth of experience and a passion for data, he is the perfect guide to take readers on a journey through the world of data analysis and machine learning. From foundational concepts to advanced techniques, this book is an invaluable resource for anyone looking to improve their data skills and take their career to the next level.

WHO THIS BOOK IS FOR?

"Python Machine Learning: A Beginner's Guide to Scikit-Learn" is intended for a wide range of readers, including:

- Individuals who are new to the field of machine learning and want to gain a solid understanding of the basics and how to apply them using the popular scikit-learn library in Python.
- Data scientists, statisticians, and analysts who are familiar with machine learning concepts but want to learn how to implement them using Python and scikit-learn.
- Developers and engineers who want to add machine learning to their skill set and build intelligent applications using Python.
- Students and researchers who are studying machine learning and want to learn how to apply it using a widely used and accessible library like scikit-learn.

The book is designed to be accessible to readers with little to no programming or math background, but still provides enough detail for more advanced readers to deepen their understanding and apply the concepts to more complex problems. The book uses a hands-on approach, with numerous code examples and practical exercises to help readers quickly learn and apply the concepts.

WHAT ARE THE REQUIREMENTS? (PRE-REQUISITES)

"Python Machine Learning: A Beginner's Guide to Scikit-Learn" is designed for individuals who have a basic understanding of Python programming and are interested in learning about machine learning. The book is ideal for students, developers, and data scientists who want to learn about machine learning in an easy-to-understand and practical way.

The following are the pre-requisites for this book:

- Basic understanding of Python programming.
- Familiarity with basic mathematical concepts such as probability and statistics is helpful but not required.
- Basic understanding of computer science concepts such as algorithms and data structures is helpful but not required.
- Basic understanding of machine learning concepts is helpful but not required.
- Access to a computer with Python and the scikit-learn library installed.

It is recommended that readers have some experience with Python and some understanding of statistics, but no prior experience with machine learning is required. The book will provide a comprehensive introduction to the scikit-learn library and the concepts of machine learning, and will guide readers through the process of building, training, and evaluating machine learning models.

PREFACE

As a data science and machine learning enthusiast, I believe that Python is one of the best programming languages for machine learning, and Scikit-Learn is one of the most powerful and user-friendly libraries for building machine learning applications.

I have written this book with the goal of providing a clear and concise introduction to the fundamentals of machine learning using Python and Scikit-Learn. Whether you are a complete beginner to machine learning or an experienced practitioner looking to learn more about using Python and Scikit-Learn, this book has something to offer you.

My hope is that by the end of this book, you will have a solid foundation in the fundamentals of machine learning, as well as the skills and knowledge to start building your own intelligent applications.

To achieve this, I have structured the book in a way that provides a step-by-step approach to learning machine learning with Python and Scikit-Learn. I start by introducing the basics of machine learning and the key concepts that you need to understand to build effective models. From there, I move on to teaching you how to use Python and Scikit-Learn to preprocess data, build and evaluate models, and improve their performance.

In addition, I have included plenty of code examples and real-world applications to help you understand how machine learning works in practice. You will also have access to a range of exercises and quizzes to help you test your understanding of the concepts covered in the book.

In addition to the core concepts and techniques of machine learning, the book also covers important aspects such as data preprocessing, feature engineering, and model deployment. It also includes practical examples and real-world use cases to help you understand how machine learning can be applied in various fields such as healthcare, finance, and robotics.

I believe that this book can help you achieve your goals in machine learning, whether it is to get started in a new career, build intelligent applications, or simply learn more about this fascinating field.

Thank you for choosing "Python Machine Learning: A Beginner's Guide to Scikit-Learn." I hope you find this book to be informative, engaging, and most importantly, helpful in your journey to becoming a skilled and knowledgeable practitioner of machine learning.

WHY SHOULD YOU READ THIS BOOK?

Are you curious about the power of machine learning and how it can be used to solve real-world problems?

Are you eager to dive into the world of Python machine learning and discover the endless possibilities it has to offer?

If so, then "Python Machine Learning: A Beginner's Guide to Scikit-Learn" is the perfect book for you. This comprehensive guide is designed to take you on a journey through the basics of machine learning and introduce you to the powerful tools and techniques available in Python.

With this book, you will learn the fundamentals of machine learning, including the concepts of supervised and unsupervised learning, and how to apply them to real-world problems. You will also discover the world of Python machine learning through hands-on examples and coding exercises. Whether you are new to machine learning or looking to expand your skills, this book will provide you with the knowledge and skills you need to start solving problems and making predictions with Python.

So, if you're ready to take your machine learning skills to the next level and explore the exciting world of Python, then pick up your copy of "Python Machine Learning: A Beginner's Guide to Scikit-Learn" today and discover the endless possibilities of this powerful tool!

Rajender Kumar

PYTHON MACHINE LEARNING: A BEGINNER'S GUIDE TO SCIKIT-LEARN

Contents

01

1 INTRODUCTION TO MACHINE LEARNING

Machine learning is a rapidly growing field that involves the use of algorithms and statistical models to analyze and make predictions or decisions based on data. This chapter will provide a background on the history and evolution of machine learning, as well as an overview of its different types and applications. Additionally, this chapter will introduce Python and scikit-learn, the popular machine learning library that will be used throughout the book. The goal of this chapter is to give readers a strong foundation in machine learning concepts and terminology, as well as the tools and techniques used in the field. By the end of this chapter, readers will have a clear understanding of the importance and potential of machine learning, and be ready to begin exploring the various algorithms and techniques used in the field.

1.1 BACKGROUND ON MACHINE LEARNING

Machine learning is a subfield of artificial intelligence (AI) and is a powerful tool for solving complex problems in a variety of industries, including finance, healthcare, transportation, and more.

The history of machine learning can be traced back to the 1940s and 1950s, when researchers first began exploring the use of computers for problem-solving and decision-making. In the early days of machine learning, algorithms were primarily used for simple tasks, such as classification and clustering. However, as technology advanced and data became more readily available, machine learning began to evolve and expand into more complex applications.

One of the key milestones in the history of machine learning was the development of the perceptron algorithm in the 1950s. The perceptron was the first algorithm capable of learning from data and was used for simple pattern recognition tasks. This was followed by the development of other algorithms, such as decision trees and artificial neural networks, which further expanded the capabilities of machine learning.

In the 1980s and 1990s, machine learning began to gain more widespread acceptance, with the development of more sophisticated algorithms and the increasing availability of data. The introduction of big data, powerful computing resources and more advanced algorithms such as Random Forest, Gradient Boosting Machine and Support Vector Machine (SVM) has led to the current state of machine learning where it is applied in a wide range of industries to solve complex problems.

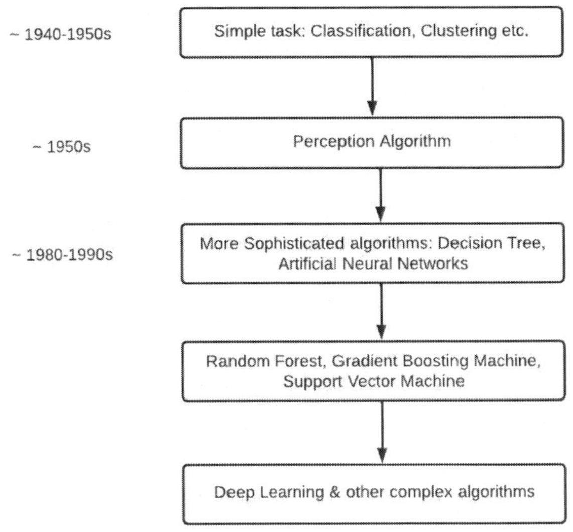

There are several different types of machine learning, including supervised learning, unsupervised learning, and deep learning. Supervised learning involves using labeled data to train a model, which can then be used to make predictions on new, unseen data. Unsupervised learning, on the other hand, involves using unlabeled data to identify patterns or structures in the data. Deep learning, a subset of machine learning, uses artificial neural networks to analyze large amounts of data and make predictions or decisions.

Machine learning is a powerful tool for solving complex problems, but it is not without its limitations. One of the main challenges of machine learning is dealing with large amounts of data, which can be difficult to process and analyze. Additionally, machine learning models can be prone to overfitting and underfitting, which can lead to inaccurate predictions or decisions. Despite these limitations, machine learning has the potential to revolutionize a wide range of industries and has already been used to solve problems that were once thought to be impossible.

In conclusion, Machine Learning is a field of computer science that uses statistical models and algorithms to analyze and make predictions or decisions based on data. Its history can be traced back to the 1940s and 1950s, and has evolved over time with the development of more sophisticated algorithms and the increasing availability of data. With the power of big data, powerful computing resources and more advanced algorithms, machine learning has become a powerful tool for solving complex problems in a wide range of industries.

1.2 WHY PYTHON FOR MACHINE LEARNING

Python is a high-level programming language that is widely used in the field of machine learning. It is an open-source language, which means it is free to use and can be modified by anyone. Python's popularity has grown rapidly in recent years, due to its ease of use, readability, and versatility.

Python is one of the most popular programming languages for machine learning, and for good reason. It offers a wide range of powerful libraries and frameworks that make it easy to implement machine learning algorithms and preprocess data. In this section, we will discuss some of the reasons why Python is the go-to language for machine learning.

Ease of Use

One of the main reasons why Python is popular for machine learning is its ease of use. The language has a simple, easy-to-read syntax that makes it easy to write and understand code. Additionally, Python has a large and active community, which means that there are many resources available to help with any problems or questions that may arise.

Powerful Libraries and Frameworks

Python has a wide range of powerful libraries and frameworks that make it easy to implement machine learning algorithms and preprocess data. Some of the most popular libraries and frameworks include:

- **Scikit-learn**: A popular library for machine learning that provides a wide range of algorithms, including linear regression, decision trees, and k-means clustering.
- **TensorFlow**: A powerful library for deep learning that makes it easy to build, train, and deploy neural networks.
- **Keras:** A high-level library for deep learning that can be used with TensorFlow and other backends.

- **Pandas**: A library for data manipulation and analysis that makes it easy to work with structured data.
- **NumPy**: A library for numerical computation that provides support for large, multi-dimensional arrays and matrices.

These libraries and frameworks make it easy to implement machine learning algorithms and preprocess data, which means that developers can focus on the problem they are trying to solve, rather than the details of the implementation.

Support for Machine Learning

Python has a large and active community that is dedicated to machine learning. This means that there are many resources available to help with any problems or questions that may arise. Additionally, there are many tutorials, books, and online courses that can help developers learn how to use Python for machine learning.

Support for Big Data

Python also has a wide range of libraries and frameworks that make it easy to work with big data, such as PySpark, Dask, and Pandas. These libraries make it easy to work with large datasets, which is important for machine learning, as the more data that is available, the better the model can perform.

Python's versatility is also one of its key advantages. It can be used for a wide range of tasks, including web development, data analysis, and scientific computing. Additionally, Python has strong support for data visualization, which is important for understanding and interpreting machine learning models.

Despite its many advantages, Python is not without its limitations. One of the main disadvantages is that it can be slow for certain tasks, such as image processing and other computationally intensive tasks. Additionally, Python is a dynamically typed language, which can make it more difficult to debug and optimize code.

In conclusion, Python is a popular choice for machine learning because it is easy to use, has a wide range of powerful libraries and frameworks, has a large and active community, and supports big data. Additionally, it has a simple and easy-to-read syntax, which makes it easy to write and understand code, which is important in machine learning because it makes it easier to experiment with different models and algorithms. It also has a wide range of resources available which makes it easy for developers to learn the language and its machine learning libraries.

1.3 OVERVIEW OF SCIKIT-LEARN

Scikit-learn is a popular machine learning library for Python. It is an open-source library, which means it is free to use and can be modified by anyone. Scikit-learn provides a wide range of tools and algorithms for building and training machine learning models, making it a powerful and versatile library. Here in the below section, we will provide some of the advantages and limitation of scikit-learn:

Advantage of scikit-learn

Here are some of the advantages of using scikit-learn:

1. **Easy to use**: Scikit-learn is designed to be easy to use, even for beginners who are just starting out in machine learning. It provides a simple and consistent interface for building and evaluating models, which can help save time and reduce errors.

2. **Comprehensive**: Scikit-learn provides a comprehensive set of tools for data preprocessing, feature selection, model selection, and evaluation. It includes a wide range of machine learning algorithms, including linear regression, logistic regression, decision trees, random forests, support vector machines, and neural networks, among others.

3. **Open source**: Scikit-learn is an open-source library, which means that it is free to use and can be customized and modified to suit your specific needs. It is also constantly being updated and improved by a large community of developers and researchers.

4. **Fast and scalable**: Scikit-learn is designed to be fast and scalable, even for large datasets. It can run on multiple cores and supports distributed computing, which makes it suitable for use in production environments.

5. **Interoperable**: Scikit-learn is interoperable with other libraries and tools, including NumPy, Pandas, and Matplotlib, which makes it easy to integrate into your existing data analysis and machine learning workflows.

6. **Extensible**: Scikit-learn is highly extensible, which means that you can add your own custom models, algorithms, and metrics to the library. This allows you to tailor the library to your specific needs and use cases.

7. **Well documented**: Scikit-learn is well documented and provides a wide range of examples and tutorials, which can help you get up and running quickly. It also has an active community forum where you can ask questions and get help from other users.

Overall, scikit-learn is a powerful and versatile machine learning library that provides a range of tools for data analysis and predictive modeling. Its ease of use, comprehensive set of tools, and open-source nature make it a popular choice for machine learning practitioners and researchers.

Limitation of scikit-learn

While it has many strengths and is widely used in the data science community, there are also some limitations to consider.

1. **Limited Deep Learning Capabilities**: Scikit-learn is not designed for deep learning, which is an area of machine learning that focuses on neural networks with many layers. While the library has some neural network functionality, it is not as extensive as other deep learning frameworks like TensorFlow or PyTorch.

2. **Lack of Support for Streaming Data**: Scikit-learn is primarily designed for batch learning, which means that it works well with datasets that can fit into memory. It does not provide support for streaming data, which is a scenario where data is continuously generated and must be processed in real-time.

3. **Limited Support for Unstructured Data**: Scikit-learn is designed for structured data, which means that it is not well-suited to dealing with unstructured data like text or images. While there are some tools for text analysis in the library, they are not as extensive as those available in specialized NLP (Natural Language Processing) libraries.

4. **Limited Parallelism**: While scikit-learn does support parallel processing, it is not optimized for distributed computing. This means that it can be slow to process very large datasets or to run complex models on large clusters.

5. **Limited AutoML Capabilities**: Scikit-learn does not have built-in tools for automating machine learning workflows, such as hyperparameter tuning or feature engineering. While some third-party libraries like TPOT provide this functionality, it is not as integrated as in other AutoML platforms.

Despite these limitations, scikit-learn is still a powerful tool for many machine learning tasks and is widely used in the data science community. It is important to understand its strengths and weaknesses when choosing a machine learning library for a given project.

1.4 SETTING UP THE DEVELOPMENT ENVIRONMENT

A development environment is a set of tools and software that allow you to write, test, and debug code. In this section, we will discuss how to set up a development environment for working with Python and machine learning.

Installing Python

The first step in setting up a development environment is to install Python. Python can be downloaded from the official website (python.org) and can be installed on Windows, macOS, or Linux.

In this section, we will discuss the step-by-step process for installing Python on each operating system, along with screenshots.

For Windows:

1. Go to the official Python website (https://www.python.org/).
2. Click on the "Downloads" button.

3. Click on the "Windows" button.
4. Click on the "Latest Python 3 Release" button. For example, it is “Python 3.11.12” at this moment.
5. Click on the "Windows x86-64 executable installer" button to download the installer.
6. Once the download is complete, double-click on the installer file to begin the installation process.
7. On the first screen of the installer, click on the "Install Now" button.

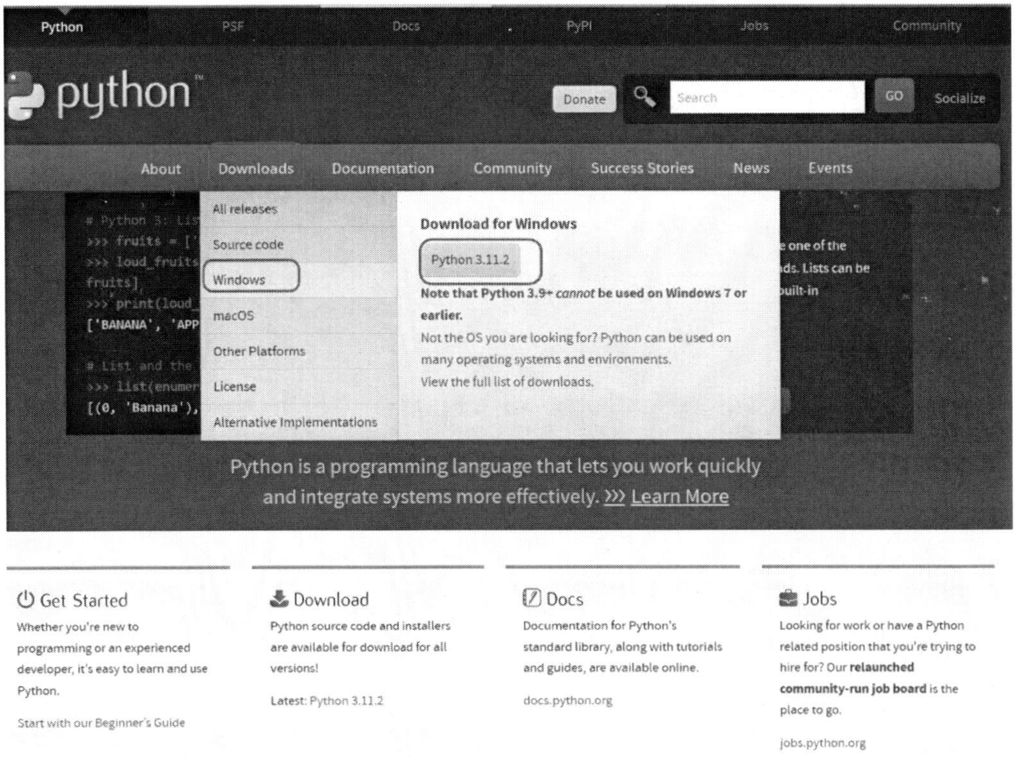

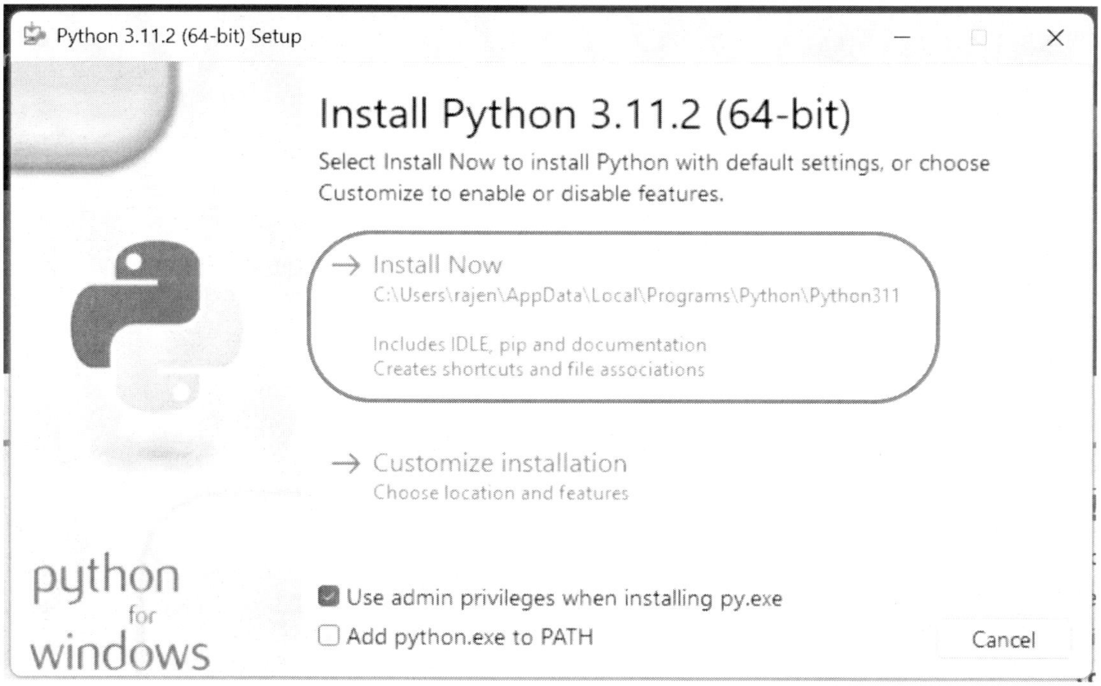

8. On the next screen, you will be asked to confirm the installation location. Leave the default location selected and click on the "Next" button.

9. On the next screen, you will be asked if you want to add Python to your system's PATH. It is recommended to select "Add Python 3.x to PATH" and click on the "Install" button.

10. Once the installation is complete, click on the "Close" button.

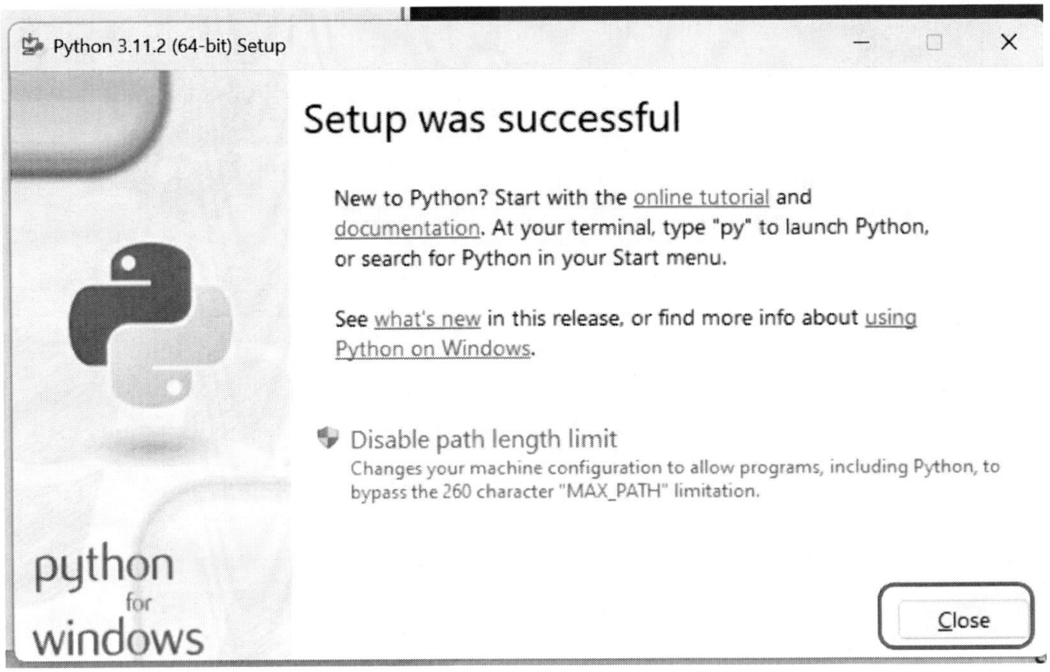

For MacOS:

1. Go to the official Python website (https://www.python.org/
2. Click on the "Downloads" button.
3. Click on the "MacOS" button.
4. Click on the "Latest Python 3 Release" button.

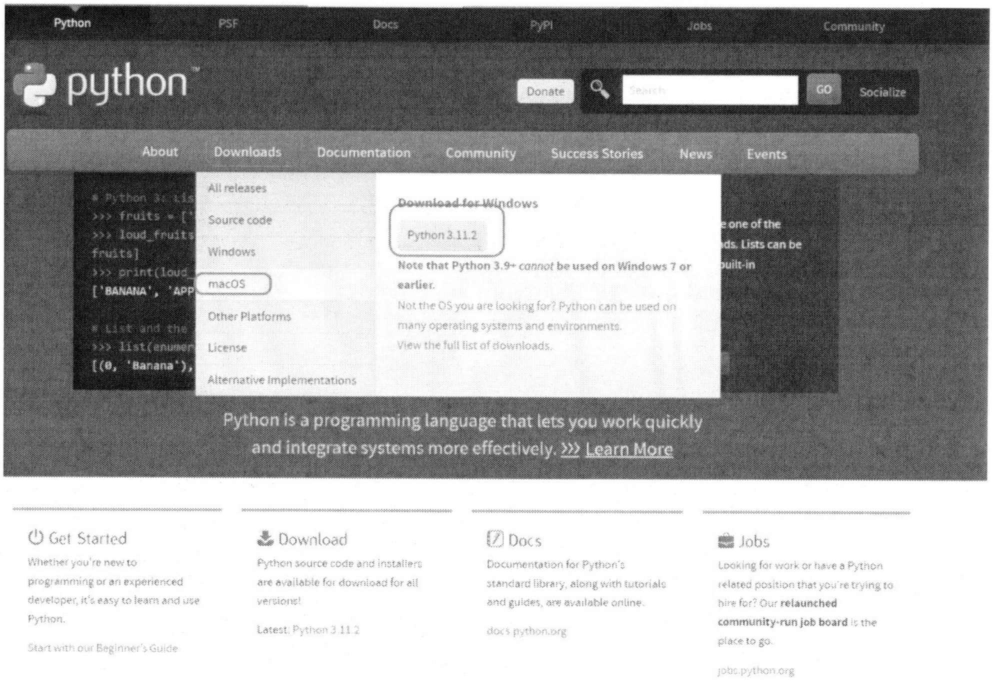

5. Click on the "macOS 64-bit installer" button to download the installer.
6. Once the download is complete, double-click on the installer file to begin the installation process.
7. On the first screen of the installer, click on the "Continue" button.
8. On the next screen, you will be asked to confirm the installation location. Leave the default location selected and click on the "Install" button.
9. Once the installation is complete, click on the "Close" button.

For Linux:

1. Open a terminal window.

2. Run the command "*sudo apt-get update*" to update your system's package list.

3. Run the command "*sudo apt-get install python3*" to install Python.

4. To check if Python is installed correctly, run the command "*python3 -V*". This should display the version of Python that is currently installed on your system.

It is important to note that the version of Python you have installed is important because some libraries and frameworks may only be compatible with specific versions of Python. It is recommended to install latest stable version of Python available.

Once Python is installed, it is recommended to install a package manager such as pip, which will make it easier to install and manage libraries and frameworks.

Installing Libraries & Frameworks

The next step is to install the libraries and frameworks that will be used in the development environment. The most popular libraries and frameworks for machine learning include NumPy, SciPy, and scikit-learn. These libraries can be installed using pip by running the command "*pip install numpy scipy scikit-learn*" on the command line.

In this section, we will discuss the step-by-step process for installing the most popular libraries and frameworks for machine learning, such as NumPy, SciPy, and scikit-learn.

The first step in installing libraries and frameworks is to make sure that Python is installed on your system. Once Python is installed, you can use the package manager pip to install libraries and frameworks.

The process of installing libraries and frameworks using pip is as follows:

1. Open a terminal or command prompt window
2. Run the command "*pip install numpy*" to install the NumPy library
3. Run the command "*pip install scipy*" to install the SciPy library
4. Run the command "*pip install scikit-learn*" to install the scikit-learn library

It is important to note that the above commands will install the latest version of the libraries and frameworks. If you need to install a specific version, you can use the command "*pip install numpy==x.x.x*" (where x.x.x is the version number)

Another way to install these libraries and frameworks is by using the Anaconda distribution which comes with a lot of useful libraries and frameworks pre-installed, and it also has a built-in package manager called Conda. You can install these libraries by running the following commands in the Anaconda prompt:

1. Open the Anaconda prompt
2. Run the command "*conda install numpy*" to install the NumPy library
3. Run the command "*conda install scipy*" to install the SciPy library
4. Run the command "*conda install scikit-learn*" to install the scikit-learn library

Once the libraries and frameworks are installed, you can import them in your Python script and start using them for building and training machine learning models.

It's important to note that, even though the process of installing libraries and frameworks is simple, it's also important to keep them updated. This is because new versions of libraries and frameworks may have bug fixes, performance improvements, or new features. You can update the libraries and frameworks by running the following commands:

1. Open a terminal or command prompt window
2. Run the command "*pip install --upgrade numpy*" to update the NumPy library
3. Run the command "*pip install --upgrade scipy*" to update the SciPy library
4. Run the command "*pip install --upgrade scikit-learn*" to update the scikit-learn library

It is also recommended to install a code editor or integrated development environment (IDE) such as Jupyter Notebook, Spyder, or PyCharm. These tools provide a user-friendly interface for writing, testing, and debugging code. They also provide features such as code completion and debugging tools that can make the development process more efficient.

To further enhance the development process, it is also recommended to have a version control system (VCS) such as Git, which allows you to keep track of changes to your code and collaborate with other developers.

In addition to these tools, it is also a good idea to have access to a large dataset that can be used to train and test machine learning models. There are many publicly available datasets that can be used for machine learning, such as the UCI Machine Learning Repository, which provides a wide range of datasets for classification, regression, and clustering tasks. You can learn more about free data available online in Appendix D at the end of book.

Setting up a development environment is an important step in working with machine learning using Python. A development environment includes tools such as Python, pip, libraries and frameworks, a code editor or IDE, version control system and datasets. By having all these tools in place, it will make the development process more efficient and streamlined.

1.5 UNDERSTANDING THE DATASET

Understanding the dataset is a crucial step in the machine learning process. It involves gaining a deep understanding of the data that will be used to train and test a model, including its structure, quality, and characteristics. This understanding is essential for selecting the appropriate model, developing a robust algorithm, and interpreting the results. In this section, we will discuss the importance of understanding the dataset, and the key considerations when working with a dataset.

The Importance of Understanding the Dataset

Before a model can be trained and tested, the dataset must be understood. This is because the dataset is the foundation upon which the model will be built, and a poor understanding of the dataset can lead to poor model performance. For example, if the dataset is not representative of the problem or is biased, the model will not perform well. Similarly, if the dataset is too small or has missing data, the model will be under-trained and will not generalize well to new data.

Understanding the dataset also helps to identify potential issues such as outliers, missing data, and duplicate records, which can affect the performance of the model. By identifying these issues early, they can be addressed before the model is trained, resulting in a more robust model.

Key Considerations when Working with a Dataset

When working with a dataset, there are several key considerations to keep in mind:

- **Representativeness**: The dataset should be representative of the problem being solved, otherwise the model may not perform well on new data.
- **Size**: The size of the dataset will affect the performance of the model. A larger dataset can result in a more robust model, but it can also increase the computational resources required to train and test the model.
- **Quality**: The quality of the data can affect the performance of the model. Missing data, outliers, and duplicate records can all affect the performance of the model.
- **Features**: The features of the data should be carefully selected, as they will be used to train and test the model. The features should be relevant to the problem and should not include redundant information.
- **Preprocessing**: The data may need to be preprocessed before it can be used to train and test the model. This can include cleaning, normalizing, and transforming the data.

In conclusion, understanding the dataset is a crucial step in the machine learning process. It involves gaining a deep understanding of the data, including its structure, quality, and characteristics. This understanding is essential for selecting the appropriate model, developing a robust algorithm, and interpreting the results. By understanding the types of data, potential issues, and key considerations when working with a dataset, machine learning practitioners can ensure that their models are robust, accurate, and generalize well to new data.

1.6 TYPE OF DATA

When working with datasets in machine learning, it is important to understand the different types of data that can be encountered. Data can be broadly classified into two categories: structured and unstructured. Understanding the different types of data and their characteristics can help in selecting the appropriate machine learning algorithms and preprocessing techniques.

Structured Data

Structured data is data that is organized and can be easily understood by a computer. It can be represented in tabular form and can be easily stored in a relational database. Examples of structured data include numerical data (such as age, income) and categorical data (such as gender, occupation). Structured data can be further classified into two types: numerical and categorical.

Numerical Data: Numerical data is data that consists of numbers. It can be further divided into two types: **discrete and continuous**. Discrete data is data that can only take on certain values, such as the number of children in a family. Continuous data, on the other hand, can take on any value within a range, such as the temperature.

Categorical Data: Categorical data is data that consists of categories. Examples include gender, occupation, and country of origin. Categorical data can be further divided into two types: **ordinal and nominal**. Ordinal data is data that can be ordered, such as education level (high school, undergraduate, graduate). Nominal data is data that cannot be ordered, such as gender (male, female).

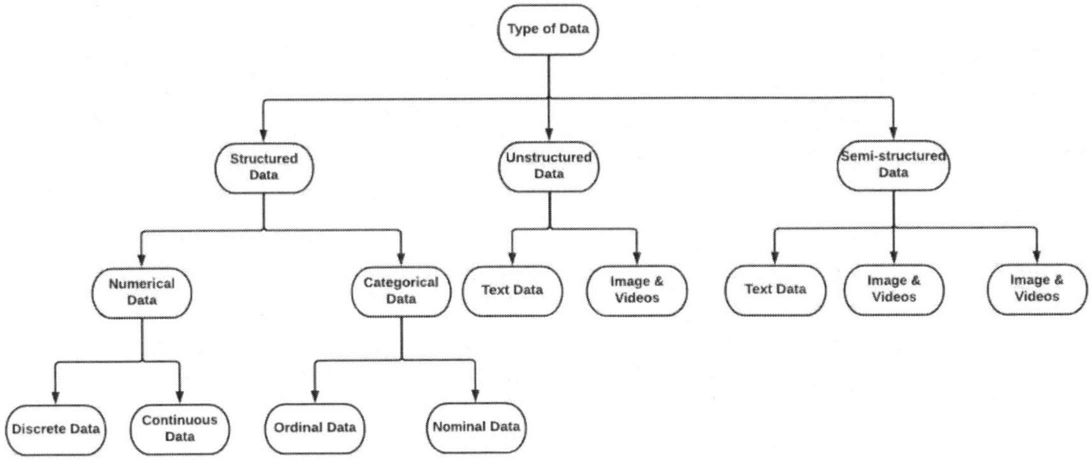

Unstructured Data

Unstructured data is data that is not organized and is difficult for a computer to understand. Examples include text, images, and audio. Unstructured data is often more challenging to work with, as it requires additional preprocessing steps to convert it into a format that can be understood by a computer. This can include text mining, image processing, and audio processing.

Text data is one of the most common types of unstructured data. It can be found in emails, social media posts, and customer reviews. Text data requires preprocessing techniques such as tokenization, stemming, and lemmatization to convert it into a format that can be understood by a machine learning model.

Images and videos are also common types of unstructured data. They require preprocessing techniques such as image processing, object detection, and feature extraction to convert them into a format that can be understood by a machine learning model.

Semi-structured Data

Semi-structured data is a type of data that does not conform to a specific data model or schema. It contains elements of both structured and unstructured data. Semi-structured data is often represented in a format that can be easily processed by computers, such as JSON, XML, or CSV files, but it may not have a fixed structure or defined data types.

One of the main advantages of semi-structured data is its flexibility. Because the data does not have to conform to a specific schema, it can be easily modified or extended to accommodate new data elements. This makes it well-suited for applications that deal with rapidly changing data or data that has a high degree of variability.

Semi-structured data is commonly used in applications such as web scraping, social media analysis, and machine learning. It is also often used as an intermediary format for converting data between different systems or applications.

However, working with semi-structured data can also have some drawbacks. Because the data is not well-defined, it can be more difficult to perform certain types of analysis, such as querying or joining data across multiple sources. It may also require more preprocessing and cleaning before it can be used in a particular application.

An **example** of working with different types of data in machine learning is building a model to predict the price of a house based on various features. Let's assume that the dataset contains the following features:

- Sale Price (Target variable)
- Number of Bedrooms (Numerical Data)

- Number of Bathrooms (Numerical Data)
- Size of the House in square feet (Numerical Data)
- Type of the House (Categorical Data, Nominal)
- Year Built (Numerical Data)
- Zipcode (Categorical Data, Nominal)

In this example, the target variable is the Sale Price, which is a numerical data as it is represented by a number, and we want to predict this variable based on other variables.

The first feature is the number of bedrooms, which is numerical data as it is represented by a number. This feature can be used as is in the model, and it can be useful in predicting the Sale Price as the number of bedrooms can affect the overall size of the house and the price.

The second feature is the number of bathrooms, which is also numerical data as it is represented by a number. This feature can also be used as is in the model, and it can be useful in predicting the Sale Price as the number of bathrooms can affect the overall amenities of the house and the price.

The third feature is the size of the house in square feet, which is numerical data as it is represented by a number. This feature can also be used as is in the model, and it can be useful in predicting the Sale Price as the size of the house can affect the overall space of the house and the price.

The fourth feature is the type of the house, which is categorical data, nominal type, as it consists of categories such as "Single Family", "Townhouse", "Apartment" etc. This feature can not be used as is in the model, as the model will not be able to understand the categorical data. So, we will use one-hot encoding to convert this feature into numerical data. One-hot encoding creates a new binary column for each category and assigns a value of 1 or 0 depending on whether the category is present in the original data or not.

The fifth feature is the year built, which is numerical data as it is represented by a number. This feature can also be used as is in the model, and it can be useful in predicting the Sale Price as the age of the house can affect the overall condition of the house and the price.

The sixth feature is the zipcode, which is also categorical data, nominal type, as it consists of categories such as "90210", "10001" etc. Similar to the type of the house, this feature also needs to be transformed using one-hot encoding.

Once the data is preprocessed, it can be used to train and test a machine learning model, such as a linear regression or a decision tree, to predict the Sale Price based on the other features. By understanding the types of data and their characteristics, we were able to preprocess the data and feed it into a model for further analysis.

In this example, we have covered three types of data: numerical, categorical (nominal), and unstructured. Understanding the types of data and preprocessing them accordingly is an important step in the machine learning process, as it ensures that the data is in a format that can be understood by a machine learning model, and that it accurately represents the problem we are trying to solve.

Understanding the different types of data and their characteristics is an important step in the machine learning process. It can help in selecting the appropriate machine learning algorithms and preprocessing techniques, and in understanding the potential issues and challenges that can be encountered when working with a dataset. Structured data is easier to work with as it is organized and can be easily understood by a computer. Unstructured data is more challenging as it requires additional preprocessing steps to convert it into a format that can be understood by a machine learning model.

1.7 TYPES OF MACHINE LEARNING MODELS

Machine learning models are algorithms that are used to make predictions or take decisions based on data. There are several types of machine learning models, each with their own strengths and weaknesses. Understanding the different types of models is crucial for selecting the right model for a specific problem and for interpreting the results of a model. The main types of machine learning models are:

1. **Supervised learning**: Supervised learning models are used to make predictions based on labeled data. The model is trained on a labeled dataset, where the output variable is known. Once the model is trained, it can be used to make predictions on new, unseen data. Examples of supervised learning models include linear regression, logistic regression, and decision trees.

2. **Unsupervised learning**: Unsupervised learning models are used to find patterns or structure in unlabeled data. The model is not given any labeled data, and instead must find patterns and structure on its own. Examples of unsupervised learning models include k-means clustering, hierarchical clustering, and principal component analysis.

3. **Semi-supervised learning**: Semi-supervised learning models are a combination of supervised and unsupervised learning. The model is given some labeled data, but not enough to fully train the model. The model must use the labeled data and the structure of the unlabeled data to make

predictions. Examples of semi-supervised learning models include self-training and co-training.

4. **Reinforcement learning**: Reinforcement learning models are used to make decisions in an environment where the model is given feedback in the form of rewards or penalties. The model learns to make decisions by maximizing the rewards over time. Examples of reinforcement learning models include Q-learning and SARSA.

5. **Deep Learning**: Deep learning models are a subfield of machine learning that is inspired by the structure and function of the human brain. These models are composed of multiple layers of interconnected nodes, called artificial neurons, which can learn and represent highly complex patterns in data. Examples of deep learning models include convolutional neural networks, recurrent neural networks, and deep belief networks.

In conclusion, understanding the different types of machine learning models is crucial for selecting the right model for a specific problem and for interpreting the results of a model. Each model has its own strengths and weaknesses, so it is important to evaluate the suitability of each model for a specific task.

We will discuss different models in the next few chapters in detail.

1.8 SUMMARY

- Introduction to Machine Learning and why Python is a popular choice for ML
- Setting up the environment by installing Python and required libraries, including a tutorial on Jupyter Notebook

- Understanding Python data structures and the importance of understanding the dataset before starting the ML process
- Introduction to Scikit-learn, a popular machine learning library in Python
- Understanding the API of scikit-learn and how it can be used to train and test models.

1.9 TEST YOUR KNOWLEDGE

I. What is the main goal of machine learning?

a) To understand and analyze data
b) To make predictions or decisions
c) To automate tasks
d) All of the above

II. What is the main advantage of using Python for machine learning?

a) It has a simple, easy-to-read syntax
b) It has a wide range of powerful libraries and frameworks
c) It has a large and active community
d) All of the above

III. What is Jupyter Notebook?

a) A library for data manipulation and analysis
b) A web-based interactive development environment
c) A machine learning algorithm
d) A database management system

IV. What is the difference between structured and unstructured data?

a) Structured data is organized and can be easily understood by a computer, while unstructured data is not organized and is difficult for a computer to understand
b) Structured data is numerical, while unstructured data is categorical
c) Structured data is easily stored in a relational database, while unstructured data is not
d) All of the above

V. **What is Exploratory Data Analysis (EDA)?**

a) A machine learning algorithm
b) A technique used to understand and analyze data
c) A preprocessing step for unstructured data
d) A data visualization library

VI. **What is one-hot encoding?**

a) A technique to convert categorical data into numerical data
b) A technique to convert numerical data into categorical data
c) A technique to remove outliers from the data
d) A technique to normalize the data

VII. **What is the main benefit of preprocessing the data?**

a) It makes the data easier to understand
b) It makes the data more representative of the problem
c) It improves the performance of the machine learning model
d) All of the above

VIII. **What is the main disadvantage of using unstructured data?**

a) It is not organized and is difficult for a computer to understand
b) It requires additional preprocessing steps to convert it into a format that can be understood by a machine learning model
c) It is not as easily stored in a relational database
d) All of the above

IX. **What is the main benefit of using scikit-learn?**

a) It provides a wide range of machine learning algorithms
b) It is easy to use and understand
c) It has a large and active community
d) All of the above

X. What is the main importance of evaluating model performance?

a) It helps to determine the accuracy of the model
b) It helps to identify areas for improvement
c) It helps to compare the performance of different models
d) All of the above

1.10 ANSWERS

I.	Answer: b)	To make predictions or decisions
II.	Answer: d)	All of the above
III.	Answer: b)	A web-based interactive development environment
IV.	Answer: a)	Structured data is organized and can be easily understood by a computer, while unstructured data is not organized and is difficult for a computer to understand
V.	Answer: b)	A technique used to understand and analyze data
VI.	Answer: a)	A technique to convert categorical data into numerical data
VII.	Answer: d)	All of the above
VIII.	Answer: d)	All of the above
IX.	Answer: d)	All of the above
X.	Answer: d)	All of the above

02

2 PYTHON: A BEGINNER'S OVERVIEW

Welcome to the chapter on Python: A Beginner's Overview. This chapter is designed to give you a solid foundation in the Python programming language. Python is a versatile and widely-used programming language that is perfect for beginners and experts alike. It is known for its simplicity, readability, and ease of use, making it an ideal choice for a wide range of applications. In this chapter, we will cover the basics of Python, including its uses, and basic syntax. We will also delve into more advanced topics such as control flow, functions, and working with data. By the end of this chapter, you will have a good understanding of the basics of Python programming and be well-equipped to continue learning and exploring the language.

2.1 PYTHON BASICS

Python is a powerful and versatile programming language that is widely used in a variety of industries, from web development to data science. It is known for its simple syntax, easy readability, and vast ecosystem of libraries and frameworks. If you're new to programming or are looking to learn Python, this section will cover the basics of the language, including its syntax, comments, and variables.

Syntax of Python

The syntax of Python is designed to be easy to read and understand, with a focus on readability and simplicity. Python uses **indentation** to indicate code blocks, rather than curly braces or keywords like "begin" and "end." This makes the code more organized and easy to read. Python also uses a simple, consistent syntax for basic operations like assignment and comparison. For example, the assignment operator is the single equal sign (=), and the comparison operator is the double equal sign (==).

Comments in Python

Comments in Python are used to add notes and explanations to your code, making it easier to understand and maintain. They are ignored by the interpreter and do not affect the execution of the program.

In Python, comments start with a hash symbol (#) and continue until the end of the line. For example, the following line is a comment:

```
# This is a comment
```

You can also place comments at the end of a line of code, after the statement:

```
x = 5  # This is a comment
```

Additionally, Python supports multi-line comments, which are denoted by triple quotes (either single or double). For example, the following code demonstrates a multi-line comment:

```
"""
This is a
multi-line comment
"""
```

Multi-line comments are often used to add documentation to a module, class, or function, and are also known as **docstrings**.

It is considered a best practice to include comments in your code, especially for complex or non-obvious sections of code. Comments should be clear, concise, and informative, and should be used to explain the purpose and function of the code.

It's also important to keep comments up-to-date, and delete or update them when the code changes. Comments that are no longer relevant or accurate can be confusing and misleading.

Indentation in Python

In Python, indentation is used to indicate code blocks. This means that the amount of whitespace at the beginning of a line is used to determine the level of nesting for a block of code. For example, in the following code, the statements in the if block are indented four spaces to the right of the if statement:

```
if x > 0:
    print("x is positive")
    x = x - 1
```

This indentation is important because it helps to make the code more organized and readable. It is also used to indicate the scope of loops, functions, and classes. The amount of indentation is not fixed and can be any multiple of spaces or tabs, as long as it is consistent within the same block of code.

It is also important to note that Python raises **IndentationError** when there is an inconsistent use of whitespaces. This means that if you use different amounts of whitespaces for different code blocks, you'll get an error. For example, if you use 4 spaces for one block and 2 spaces for another, you'll get an error.

To avoid this, it is recommended to use spaces or tabs consistently throughout your code, and use an editor that automatically converts tabs to spaces. Most popular Python IDEs like PyCharm, VSCode, and Jupyter notebook have this feature enabled by default.

In summary, indentation is an important aspect of Python's syntax, as it is used to indicate code blocks and helps to make the code more organized and readable. It is important to use consistent indentation throughout your code to avoid errors and improve readability.

Variable in Python

In Python, variables are used to store and manipulate data in a program. They are used to give a name to a value, so that the value can be referred to by its name rather than its value. Variables are declared by assigning a value to a variable name. For example, the following code declares a variable named "x" and assigns the value 5 to it:

```
x = 5
```

Python is a dynamically-typed language, which means that the type of a variable is determined at runtime. This means that you don't have to specify the type of a variable when you declare it. For example, the following code assigns a string to a variable:

```
name = "John"
```

It's also important to note that Python variable names must start with a letter or an underscore and can only contain letters, numbers, and underscores. Additionally, Python has a number of reserved words that cannot be used as variable names. These reserved words include keywords such as "if", "else", "for", "in", "and", "or", etc.

Python also supports multiple assignment, which allows you to assign values to multiple variables in a single line. For example, the following code assigns values to three variables in a single line:

```
x, y, z = 1, 2, 3
```

Additionally, Python also supports variable swapping, where the values of two variables can be swapped in a single line of code, without the need of a temporary variable. For example, the following code swaps the values of x and y:

```
x, y = y, x
```

In conclusion, Python is a powerful and versatile programming language that is easy to learn and use. Its simple syntax, easy readability, and vast ecosystem of libraries and frameworks make it a popular choice for a wide range of applications. This section has covered the basics of the language, including its syntax, comments, and variables. With a solid understanding of these concepts, you'll be well on your way to becoming a proficient Python programmer.

2.2 DATA TYPES IN PYTHON

In Python, data types are used to define the type of a variable or a value. Different data types have different properties and behaviors, and they are used to store different types of data. The most commonly used data types in Python are:

1. **Numbers**: Python has various types of numerical data types, such as int (integer), float (floating point number), and complex. Integers are whole numbers, positive or negative, and they do not have decimal points. For example:

```
x = 5  # int
y = -10  # int
```

Floating point numbers are numbers that have decimal points and they are used for representing real numbers. For example:

```
y = 3.14   # float
z = -0.5 # float
```

Complex numbers are numbers that consist of a real and an imaginary part. They are written in the form of a+bj, where a and b are real numbers and j is the imaginary unit. For example:

```
z = 3 + 4j   # complex
```

2. **Strings**: A string is a sequence of characters. They can be declared using single quotes ('') or double quotes (""). Strings are used to represent text and they are immutable, meaning they cannot be modified after they are created. For example:

```
name = "John"   # string
```

3. **Lists**: Lists are ordered sequences of items, which can be of any data type. They are enclosed in square brackets and the items are separated by commas. Lists are mutable, meaning they can be modified after they are created. For example:

```
fruits = ["apple", "banana", "orange"]   # list
```

4. **Tuples**: Tuples are similar to lists but they are immutable, meaning they cannot be modified after they are created. They are also enclosed in parentheses and the items are separated by commas. For example:

```
coordinates = (3, 4)  # tuple
```

5. **Dictionaries**: Dictionaries are used to store key-value pairs. They are enclosed in curly braces {} and the items are separated by commas. The keys must be unique and immutable. Dictionaries are also mutable, meaning they can be modified after they are created. For example:

```
person = {"name": "John", "age": 30}  # dictionary
```

6. **Booleans**: Booleans represent true or false values. They can be either True or False. They are mostly used in conditional statements and loops. For example:

```
is_valid = True # Boolean
```

It's important to note that in Python, the type of a variable or a value can be determined using the built-in function type(). For example:

```
x = 5
print(type(x))   # <class 'int'>
```

In conclusion, data types are an important aspect of Python programming as they define the type of a variable or a value, and they have different properties and behaviors. The most commonly used data types in Python are numbers (int, float, complex), strings, lists, tuples, dictionaries, and booleans. Choosing the appropriate data type for the task at hand is crucial, as it affects how the data can be used and manipulated in your program. Additionally, knowing the mutability of data types helps to make efficient use of memory and prevent unexpected behavior in your code.

2.3 CONTROL FLOW IN PYTHON

Control flow refers to the order in which statements in a program are executed. In Python, control flow is achieved through the use of statements such as if-else, for loops, and while loops. These statements allow you to control the flow of execution in your program and make decisions based on certain conditions.

The if-else statement is used to make decisions in your code. It allows you to execute a block of code only if a certain condition is true. The syntax for an if-else statement is as follows:

```
if condition:
    # execute this code block if condition is true
else:
    # execute this code block if condition is false
```

For example, the following code checks if a variable x is greater than 5, and if it is, it prints "x is greater than 5".

```
x = 7
if x > 5:
    print("x is greater than 5")
else:
    print("x is not greater than 5")
```

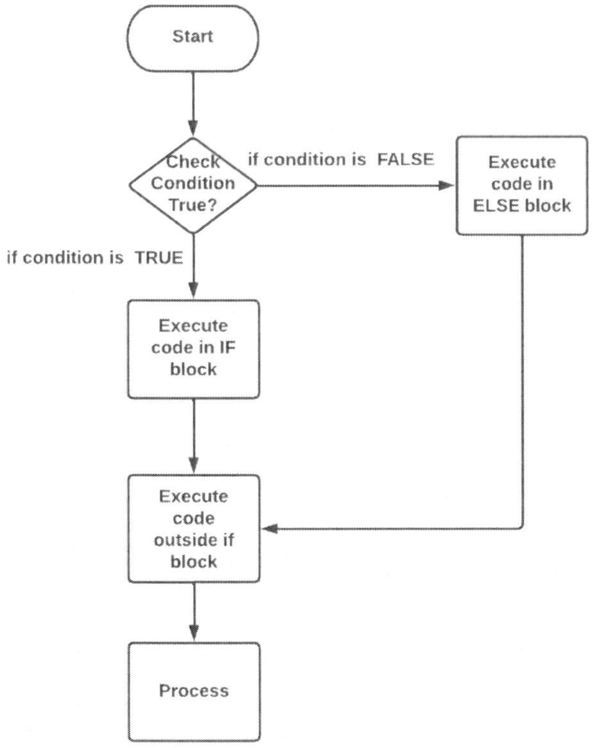

Another type of control flow structure is the **for loop**. The for loop is used to iterate through a sequence of items, such as a list, tuple, or string. The syntax for a for loop is as follows:

```
for variable in sequence:
    # execute this code block for each item in the sequence
```

For example, the following code iterates through a list of numbers, and prints each number:

```
numbers = [1, 2, 3, 4, 5]
for number in numbers:
    print(number)
```

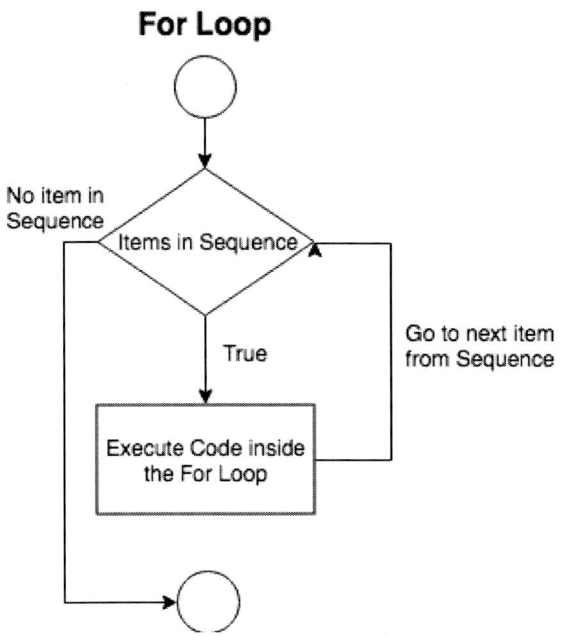

The **while loop** is another type of control flow structure. It is used to execute a block of code repeatedly as long as a certain condition is true. The syntax for a while loop is as follows:

```
while condition:
    # execute this code block while the condition is true
```

For example, the following code uses a while loop to print the numbers from 1 to 5:

```
i = 1
while i <= 5:
    print(i)
    i += 1
```

Python also provides the **break** and **continue** statements for more fine-grained control over the flow of execution within loops. The **break** statement allows you to exit a loop early, while the **continue** statement allows you to skip the current iteration and move on to the next one.

For example, the following code uses a for loop to iterate through a list of numbers, but it uses the **break** statement to exit the loop early if the number is equal to 3:

```
numbers = [1, 2, 3, 4, 5]
for number in numbers:
    if number == 3:
        break
    print(number)
```

The output of this code will be 1, 2, but not 3.

In conclusion, control flow is an important aspect of programming and Python provides a variety of tools to control the flow of execution in your program. The if-else statement, for loops, and while loops are used to make decisions and execute code repeatedly based on certain conditions. Additionally, the **break** and **continue** statements provide more fine-grained control over the flow of execution within loops. Understanding and using these control flow structures effectively will help you write more efficient and organized code.

2.4 FUNCTIO IN PYTHON

In Python, a function is a block of reusable code that performs a specific task. Functions are useful for organizing and structuring code, making it more readable and maintainable. They can also be reused across multiple parts of a program, reducing code duplication.

Functions are defined using the **def** keyword, followed by the function name, and a set of parentheses that may contain parameters. For example, the following code defines a simple function called **greet** that takes a single parameter **name**:

```
def greet(name):
    print("Hello, " + name)
```

Once a function is defined, it can be called or invoked by using its name followed by parentheses. For example:

```
greet("John")
```

This will output "Hello, John".

Functions can also return a value using the **return** statement. For example, the following function takes two parameters **a** and **b**, and returns their sum:

```
def add(a, b):
    return a + b

result = add(3, 4)
print(result)
```

This will output 7.

It's also important to note that in Python, functions can have default values for their parameters. This means that if a value is not passed for a parameter when calling the function, the default value will be used instead. For example:

```
def greet(name, message = "Hello"):
    print(message + ", " + name)

greet("John")
```

In this example, the **message** parameter has a default value of "Hello", so if it's not passed when calling the function, the default value will be used.

Functions can also be used as arguments in other functions, a technique called Higher-Order functions. This makes the code more flexible and allows the developer to write more generic and reusable functions.

In conclusion, functions are an important aspect of Python programming. They provide a way to organize and structure code, making it more readable and maintainable. Functions can also be reused across multiple parts of a program, reducing code duplication. They can also take parameters and return values.

2.5 ANONYMOUS (LAMBDA) FUNCTION

In Python, an anonymous function, also known as a lambda function, is a function without a name. They are defined using the **lambda** keyword, followed by a set of parameters, a colon, and a single expression. The expression is evaluated and returned when the function is called.

For example, the following code defines a lambda function that takes two parameters **a** and **b**, and returns their product:

```
multiply = lambda a, b: a * b
result = multiply(3, 4)
print(result)
```

This will output 12.

Lambda functions are useful when a small, simple function is needed, such as a callback function for a button click or a sorting key for a list. They can also be used as arguments in other functions, such as the **filter()** and **map()** functions.

For example, the following code uses the **filter()** function to filter a list of numbers, keeping only the even numbers:

```
numbers = [1, 2, 3, 4, 5, 6, 7, 8, 9, 10]
even_numbers = filter(lambda x: x % 2 == 0, numbers)
print(list(even_numbers))
```

It's important to note that, unlike regular functions, lambda functions do not have a return statement. The value of the expression is returned by default. Also, lambda functions cannot contain statements, only expressions.

2.6 FUNCTION FOR LIST

Python provides a variety of built-in functions that can be used to manipulate lists. These functions make it easy to perform common operations on lists, such as sorting, filtering, and mapping.

The **len()** function is used to determine the length of a list, which is the number of elements it contains. For example:

```
fruits = ['apple', 'banana', 'orange']
print(len(fruits))
```

This will output 3.

The **sort()** function is used to sort the elements of a list in ascending order. It modifies the original list and does not return a new one. For example:

```
numbers = [3, 1, 4, 2, 5]
numbers.sort()
print(numbers)
```

This will output [1, 2, 3, 4, 5].

The **sorted()** function is similar to the **sort()** function but it returns a new list rather than modifying the original one.

```
numbers = [3, 1, 4, 2, 5]
sorted_numbers = sorted(numbers)
print(sorted_numbers)
```

This will output [1, 2, 3, 4, 5].

The **filter()** function is used to filter the elements of a list based on a certain condition. It returns an iterator, which can be converted to a list. For example:

```
numbers = [1, 2, 3, 4, 5]
even_numbers = filter(lambda x: x % 2 == 0, numbers)
print(list(even_numbers))
```

This will output [2, 4].

The **map()** function is used to apply a certain operation to each element of a list. It returns an iterator, which can be converted to a list. For example:

```
numbers = [1, 2, 3, 4, 5]
squared_numbers = map(lambda x: x ** 2, numbers)
print(list(squared_numbers))
```

This will output [1, 4, 9, 16, 25].

The **reduce()** function is used to reduce a list of elements to a single value by applying a certain operation. It is a part of the **functools** module. For example:

```
from functools import reduce
numbers = [1, 2, 3, 4, 5]
product = reduce(lambda x, y: x * y, numbers)
print(product)
```

This will output 120.

In conclusion, Python provides a variety of built-in functions that make it easy to manipulate lists. The **len()**, **sort()**, **sorted()**, **filter()**, **map()** and **reduce()** functions are commonly used for tasks such as determining the length of a list, sorting elements, filtering elements based on a certain condition, applying an operation to each element of a list, and reducing a list of elements to a single value. These functions can make your code more readable, efficient, and organized. It's important to note that some of these functions such as **filter()** and **map()** returns an iterator, which can be converted to a list using the **list()** function. Additionally, **reduce()** is part of the **functools** module and needs to be imported before using it. Understanding and using these built-in functions effectively can help you write more efficient and organized code.

2.7 FUNCTION FOR DICTIONARY

Python provides a variety of built-in functions that can be used to manipulate dictionaries. These functions make it easy to perform common operations on dictionaries, such as iterating through keys and values, adding and updating key-value pairs, and checking for the existence of a key or value.

The **len()** function is used to determine the number of key-value pairs in a dictionary. For example:

```
person = {"name": "John", "age": 30}
print(len(person))
```

This will output 2.

The **keys()** function is used to return a view of all the keys in a dictionary. For example:

```
person = {"name": "John", "age": 30}
print(person.keys())
```

This will output **dict_keys(['name', 'age'])**

The **values()** function is used to return a view of all the values in a dictionary. For example:

```
person = {"name": "John", "age": 30}
print(person.values())
```

This will output **dict_values(['John', 30])**

The **items()** function is used to return a view of all the key-value pairs in a dictionary as a list of tuple. For example:

```
person = {"name": "John", "age": 30}
print(person.items())
```

This will output **dict_items([('name', 'John'), ('age', 30)])**

The **get()** function is used to retrieve the value of a key in a dictionary. If the key is not found, it returns None or a default value that can be specified as an argument. For example:

```
person = {"name": "John", "age": 30}
print(person.get("name")) # Output: John
print(person.get("address", "Unknown")) # Output: Unknown
```

The **update()** function is used to update the key-value pairs of a dictionary. It can take another dictionary or key-value pairs as argument. For example:

```
person = {"name": "John", "age": 30}
person.update({"name": "Jane", "gender": "female"})
print(person) # Output: {'name': 'Jane', 'age': 30, 'gender':
'female'}
```

The **pop()** function is used to remove a key-value pair from a dictionary. It takes the key as an argument and returns the value of the key that was removed. If the key is not found, it raises a **KeyError** or return a default value that can be specified as an argument. For example:

```
person = {"name": "John", "age": 30}
print(person.pop("name")) # Output: John
print(person) # Output: {'age': 30}
```

The **in** keyword is used to check if a key or a value exists in a dictionary. For example:

```
person = {"name": "John", "age": 30}
print("name" in person) # Output: True
print("address" in person) # Output: False
```

In conclusion, Python provides a variety of built-in functions that make it easy to manipulate dictionaries. The **len()**, **keys()**, **values()**, **items()**, **get()**, **update()**, **pop()** and the **in** keyword are commonly used for tasks such as determining the number of key-value pairs, iterating through keys and values, adding and updating key-value pairs, checking for the existence of a key or value and removing key-value pairs. These functions can make your code more readable, efficient, and organized. Understanding and using these built-in functions effectively can help you write more efficient and organized code.

2.8 STRING MANIPULATION FUNCTION

Python provides a variety of built-in functions and methods for manipulating strings. These functions and methods make it easy to perform common operations on strings, such as concatenation, slicing, formatting, and searching.

The **len()** function is used to determine the length of a string, which is the number of characters it contains. For example:

```
name = "John"
print(len(name))
```

This will output 4.

The + operator is used to concatenate two or more strings together. For example:

```
first_name = "John"
last_name = "Doe"
full_name = first_name + " " + last_name
print(full_name)
```

This will output "John Doe".

The * operator is used to repeat a string a certain number of times. For example:

```
name = "John"
print(name * 3)
```

This will output "JohnJohnJohn".

The **[:]** notation is used to slice a string. It allows you to extract a substring from a string by specifying the start and end index. For example:

```
name = "John Doe"
print(name[0:4])
```

This will output "John".

The **in** keyword is used to check if a substring exists in a string. For example:

```
name = "John Doe"
print("John" in name) # Output: True
print("Jane" in name) # Output: False
```

The **replace()** method is used to replace a substring with another substring in a string. For example:

```
name = "John Doe"
new_name = name.replace("John", "Jane")
print(new_name)
```

This will output "Jane Doe".

The **split()** method is used to split a string into a list of substrings using a specified delimiter. For example:

```
name = "John,Doe,30"
name_list = name.split(",")
print(name_list)
```

This will output **['John', 'Doe', '30']**

The **format()** method is used to format strings by replacing placeholders with values. Placeholders are represented by curly braces **{}**. For example:

```
name = "John"
age = 30
print("My name is {} and I am {} years old.".format(name, age))
```

This will output "My name is John and I am 30 years old."

The **join()** method is used to join a list of strings into a single string using a specified delimiter. For example:

```
names = ["John", "Doe", "Jane"]
string_of_names = ",".join(names)
print(string_of_names)
```

This will output "John,Doe,Jane"

In conclusion, Python provides a variety of built-in functions and methods for manipulating strings. The **len()**, + operator, * operator, **[:]** notation, **in** keyword, **replace()**, **split()**, **format()** and **join()** are commonly used for tasks such as determining the length of a string, concatenating multiple strings, repeating a string, slicing substrings, checking for the existence of a substring, replacing substrings, splitting strings into a list of substrings, formatting strings with placeholders and joining a list of strings into a single string. These functions and methods can make your code more readable, efficient, and organized. Understanding and using these built-in functions and methods effectively can help you write more efficient and organized code. It's important to note that some methods like **replace()**, **split()**, **format()** and **join()** are specific to strings and can't be used on other data types.

2.9 EXCEPTION HANDLING

Exception handling is a mechanism in Python that allows you to handle errors and exceptional situations in your code. It allows you to write code that can continue to execute even when an error occurs. This is important because it allows your program to continue running and avoid crashing, which can lead to a better user experience.

The **try** and **except** keywords are used to handle exceptions in Python. The **try** block contains the code that may raise an exception. The **except** block contains the code that will be executed if an exception is raised. For example:

```
try:
   num1 = 7
   num2 = 0
   print(num1 / num2)
except ZeroDivisionError:
   print("Division by zero is not allowed.")
```

In this example, the code in the **try** block raises a **ZeroDivisionError** exception when it attempts to divide **num1** by **num2**, which has a value of 0. The code in the **except** block is executed and the message "Division by zero is not allowed." is printed.

You can use multiple **except** blocks to handle different types of exceptions. For example:

```
try:
   variable = "hello"
   print(int(variable))
except ValueError:
   print("ValueError: could not convert string to int.")
except TypeError:
   print("TypeError: int() argument must be a string, a bytes-like
object or a number, not 'list'")
```

In this example, the code in the **try** block raises a **ValueError** exception when it attempts to convert the string "hello" to an integer, which is not possible. The first **except** block is executed and the message "ValueError: could not convert string to int." is printed.

You can also use the **finally** block to include code that will be executed regardless of whether an exception was raised or not. For example:

```
try:
   num1 = 7
   num2 = 0
   print(num1 / num2)
except ZeroDivisionError:
   print("Division by zero is not allowed.")
finally:
   print("This code will be executed no matter what.")
```

In this example, the code in the **finally** block will be executed regardless of whether the code in the **try** block raises an exception or not.

In addition, you can use the **raise** keyword to raise an exception manually.

```
raise ValueError("Invalid Value")
```

In conclusion, Exception handling is an important feature in Python that allows you to handle errors and exceptional situations in your code. The **try** and **except** keywords are used to handle exceptions, while the **finally** block can be used to include code that will be executed regardless of whether an exception was raised or not. Using multiple **except** blocks allows you to handle different types of exceptions separately. The **raise** keyword can be used to raise an exception manually. Exception handling allows your program to continue running and avoid crashing, which can lead to a better user experience. It is important to use exception handling in your code to anticipate and handle unexpected situations, and to make your code more robust and reliable.

2.10 FILE HANDLING IN PYTHON

File handling in Python allows you to read from and write to files on your computer's file system. Python provides a variety of built-in functions and methods for working with files, including opening, reading, writing, and closing files.

The **open()** function is used to open a file. It takes the file name and the mode in which the file should be opened as arguments. The mode can be 'r' for reading, 'w' for writing, and 'a' for appending. For example:

```
file = open('example.txt', 'r')
```

This opens the file 'example.txt' in read mode.

The **read()** method is used to read the contents of a file. For example:

```
file = open('example.txt', 'r')
contents = file.read()
print(contents)
file.close()
```

This reads the contents of the file 'example.txt' and stores it in the variable 'contents' and prints it. it's important to close the file after reading it by calling **file.close()**

The **write()** method is used to write to a file. It takes a string as an argument. For example:

```
file = open('example.txt', 'w')
file.write("Hello World!")
file.close()
```

This writes the string "Hello World!" to the file 'example.txt'.

The **append()** method is used to add new data to the end of a file without overwriting the existing contents. It works similar to the write method. For example:

```
file = open('example.txt', 'a')
file.write("\nThis is an added text")
file.close()
```

This will add "This is an added text" to the end of the file 'example.txt' without overwriting the existing contents.

The **with** statement is used to open a file and automatically close it when you are done with it. This is considered as a best practice to avoid file not closed errors. For example:

```
with open('example.txt', 'r') as file:
    contents = file.read()
    print(contents)
```

This will open the file 'example.txt' in read mode, read its contents and print it, and then automatically close the file when it's done.

The **readline()** method is used to read a single line from a file. For example:

```
with open('example.txt', 'r') as file:
    line = file.readline()
    print(line)
```

This will read the first line of the file 'example.txt' and print it.

File handling in Python allows you to read from and write to files on your computer's file system. Python provides a variety of built-in functions and methods for working with files, including opening, reading, writing, and closing files. It's important to close the file after reading or writing to it, and the **with** statement is considered as a best practice to avoid file not closed errors. These functions and methods can make your code more efficient and organized. Understanding and using these built-in functions and methods effectively can help you write more efficient and organized code.

2.11 MODLUES IN PYTHON

Modules in Python are pre-written code libraries that you can use to add extra functionality to your Python programs. They are a way to organize and reuse code, and they can help you write more efficient and organized code. Python has a wide variety of built-in modules, and you can also install third-party modules using package managers such as pip.

The **import** statement is used to import a module in Python. For example:

```
import math
```

This imports the math module, which provides mathematical functions such as **sqrt()**, **sin()**, and **cos()**.

You can also use the **from** keyword to import specific functions or variables from a module. For example:

```
from math import sqrt
```

This imports only the **sqrt()** function from the math module.

You can also use the **as** keyword to give a module or function a different name when you import it. For example:

```
import math as m
```

This imports the math module and gives it the name 'm', so you can use **m.sqrt()** instead of **math.sqrt()**.

You can also use the * wildcard character to import all functions and variables from a module. For example:

```
from math import *
```

This imports all functions and variables from the math module, so you can use **sqrt()** instead of **math.sqrt()**.

Python also provides a way to find out the list of functions or variable inside a module using **dir()** function. For example

```
import math
print(dir(math))
```

This will print a list of all the functions and variables inside the math module.

It's important to note that using the * wildcard character to import all functions and variables from a module can cause naming conflicts if there are functions or variables with the same name in different modules. It's recommended to use the **import** statement to import the specific functions or variables that you need, or to use the **as** keyword to give them different names.

In conclusion, modules in Python are pre-written code libraries that you can use to add extra functionality to your Python programs. They are a way to organize and reuse code, and they can help you write more efficient and organized code. The **import** statement, **from** keyword, **as** keyword, and * wildcard character are used to import modules, functions, and variables. It's important to use the import statement to import the specific functions or variables that you need and to use the **dir()** function to find out the list of functions or variables inside a module. Understanding and using modules effectively can help you write more efficient and organized code, and it can save you time by not having to re-write code that already exists.

2.12 STYLE GUIDE FOR PYTHON CODE

To make the code readable, maintainable, and shareable, there are certain style conventions that need to be followed. These conventions are documented in the Python Style Guide, also known as PEP 8.

PEP 8 provides a set of guidelines for formatting Python code. These guidelines cover topics such as naming conventions, indentation, whitespace, line length, comments, and more. By following these guidelines, you can improve the readability and consistency of your code, which makes it easier to understand and maintain.

Here are some of the key style conventions that are recommended by PEP 8:

1. **Naming Conventions**

In Python, there are naming conventions for variables, functions, classes, and modules. These conventions make it easier to understand the purpose of each object. Here are the basic naming conventions:

- Variables and functions should be lowercase, with words separated by underscores.
- Classes should use CamelCase (words are separated by capital letters).
- Modules should be lowercase, with words separated by underscores.

2. **Indentation**

Python uses indentation to define blocks of code, rather than using braces or other delimiters. It's recommended to use 4 spaces for each level of indentation, rather than using tabs.

3. **Whitespace**

Whitespace can be used to improve the readability of your code. It's recommended to use a single space after commas and operators, and to avoid using spaces between function names and parentheses.

4. **Line Length**

Long lines of code can be difficult to read, especially on smaller screens or in a terminal window. PEP 8 recommends limiting lines to a maximum of 79 characters. If a line of code exceeds this limit, you can break it up into multiple lines using backslashes or parentheses.

5. **Comments**

Comments can be used to explain the purpose of your code and how it works. PEP 8 recommends using comments sparingly, and only when necessary. Comments should be written in complete sentences, and should start with a capital letter.

In addition to these conventions, PEP 8 also recommends a few other best practices. For example, it's recommended to use the built-in functions and modules whenever possible, rather than writing your own. It's also recommended to use the Python 3.x syntax whenever possible, rather than using deprecated features from Python 2.x.

By following these conventions, you can make your Python code more readable, maintainable, and shareable. If you're working on a large project with other developers, it's especially important to follow these guidelines, as it makes it easier for everyone to understand the code.

PEP 8 provides a set of guidelines for formatting Python code. By following these guidelines, you can improve the readability and consistency of your code, which makes it easier to understand and maintain.

2.13 DOCSTRING CONVENTIONS IN PYTHON

In Python, a docstring is a string literal that appears as the first statement of a module, function, class, or method definition. It is used to provide documentation about the object being defined.

There are several conventions for writing docstrings in Python, with the most common being the PEP 257 convention. This convention specifies that a docstring should be a multi-line string that includes a one-line summary of the object, followed by a blank line and a more detailed description of the object. The detailed description should be in the form of complete sentences, and should provide information on the parameters, return value, and any exceptions that the function may raise.

Here's an example of a function definition with a docstring that follows the PEP 257 convention:

```
def add_numbers(x, y):
    """
    Add two numbers together and return the result.

    Args:
        x (int): The first number to add.
        y (int): The second number to add.

    Returns:
        int: The sum of x and y.
    """
    return x + y
```

In this example, the docstring provides a summary of the function and a detailed description of the parameters and return value. The parameter types are specified using type hints, which are optional but recommended in Python 3.

Following a consistent docstring convention can help make your code more readable and easier to understand. It also makes it easier for automated tools to generate documentation from your code.

2.14 PYTHON LIBRARY FOR DATA SCIENCE

Python is a popular programming language that has gained immense popularity in the data science community. It offers a vast array of libraries and tools that have made data science tasks easier and more efficient. In this article, we will discuss some of the most popular Python libraries for data science.

1. **NumPy**: NumPy is a fundamental library for scientific computing in Python. It provides support for large, multi-dimensional arrays and matrices, along with a large library of mathematical functions. NumPy is designed to integrate with other libraries like SciPy and Pandas.

2. **Pandas**: Pandas is a library that provides data manipulation and analysis tools. It offers data structures like Series and DataFrame that allow for easy manipulation and transformation of data. Pandas provides support for working with tabular, structured, and time-series data.

3. **Matplotlib**: Matplotlib is a 2D plotting library that enables users to create various types of charts and plots. It supports a wide range of plot types, including line, bar, scatter, and histogram. Matplotlib is an excellent tool for visualizing data and generating insights.

4. **Scikit-learn**: Scikit-learn is a machine learning library for Python. It provides simple and efficient tools for data mining and data analysis. Scikit-learn offers support for various supervised and unsupervised learning algorithms.

5. **TensorFlow**: TensorFlow is an open-source machine learning library developed by Google. It is widely used for building deep learning models. TensorFlow provides an extensive set of tools for building and training neural networks, along with a high-level Keras API for building deep learning models.

6. **PyTorch**: PyTorch is another popular open-source machine learning library that is widely used for building deep learning models. It provides support for both CPU and GPU processing and is known for its simplicity and ease of use.

7. **Keras**: Keras is a high-level neural networks API, written in Python and capable of running on top of TensorFlow, CNTK, or Theano. It is designed to enable fast experimentation with deep neural networks, and it provides a simple, consistent interface for building and training deep learning models.

8. **Seaborn**: Seaborn is a data visualization library based on Matplotlib. It provides a high-level interface for drawing informative and attractive statistical graphics. Seaborn provides support for a wide range of plot types, including heatmaps, violin plots, and categorical plots.

In conclusion, Python has become the go-to programming language for data science due to the wide range of libraries and tools available. The libraries listed above are some of the most popular Python libraries for data science and are used by data scientists and machine learning engineers worldwide.

2.15 SUMMARY

- Python is a high-level programming language that is widely used for web development, data science, and other fields.
- Python has a simple and easy-to-learn syntax, making it a popular choice for beginners.
- Python has a large and active community, which provides a wide range of resources and libraries for programmers.
- There are different versions of Python, including Python 2 and Python 3, and it is recommended to use the latest version (Python 3) as it has many improvements over the older version.
- Python supports various data types including numbers, strings, lists, and dictionaries.
- Python has built-in functions and modules that can be used for tasks such as mathematical operations, string manipulation, and file handling.

- Python also has a feature called "Control Flow" which allows the programmer to control the flow of the program, making use of the 'if' and 'else' statements, the 'for' and 'while' loops.
- Python also allows defining and using functions which can be defined using the 'def' keyword and called by their name.
- Python also has a feature called "Exception Handling" which allows the programmer to handle errors and exceptional situations in their code, making use of the 'try' and 'except' keywords.
- Python also has a wide variety of libraries for data science and machine learning, including NumPy, pandas, Matplotlib, Seaborn, scikit-learn, TensorFlow, Keras, PyTorch, SciPy, and statsmodels.

Python is a powerful and versatile programming language that can be used for a wide variety of tasks. It has a simple and easy-to-learn syntax, making it a popular choice for beginners. It also has a large and active community, which provides a wide range of resources and libraries for programmers. Understanding the basics of Python, including its data types, built-in functions and modules, control flow, and exception handling, is essential for any beginner looking to start programming with Python. Additionally, being familiar with the most popular libraries for data science and machine learning can help beginners to easily implement complex algorithms and perform advanced data analysis tasks.

2.16 TEST YOUR KNOWLEDGE

I. What is the recommended version of Python to use?

a) Python 2
b) Python 3

c) Python 4
d) Python 5

II. **What is the purpose of the try and except keywords in Python?**

a) To control the flow of the program
b) To handle errors and exceptional situations
c) To import modules
d) To define and call functions

III. **What is the purpose of the import statement in Python?**

a) To control the flow of the program
b) To handle errors and exceptional situations
c) To import modules
d) To define and call functions

IV. **Which data type in Python is used to store multiple items in a single variable?**

a) String
b) Integer
c) List
d) Tuple

V. **What is the purpose of the dir() function in Python?**

a) To control the flow of the program
b) To handle errors and exceptional situations
c) To import modules
d) To find out the list of functions or variables inside a module

VI. **What is the purpose of the open() function in Python?**

a) To open a file
b) To close a file
c) To read a file
d) To write to a file

VII. **Which library in Python is widely used for topic modeling and document similarity analysis?**

a) NumPy
b) pandas
c) Gensim
d) Matplotlib

VIII. **What is the purpose of the * wildcard character when importing modules in Python?**

a) To import all functions and variables from a module
b) To import specific functions or variables from a module
c) To give a module or function a different name when importing
d) To open a file

IX. **What is the purpose of the with statement in Python when working with files?**

a) To open a file and automatically close it when you are done with it
b) To read a file
c) To write to a file
d) To find out the list of functions or variables inside a module

X. **Which library in Python is widely used for machine learning and data science competitions?**

a) NumPy
b) pandas
c) XGBoost
d) Matplotlib

2.17 ANSWERS

I.	Answer: b)	Python 3
II.	Answer: b)	To handle errors and exceptional situations
III.	Answer: c)	To import modules
IV.	Answer: c)	List
V.	Answer: d)	To find out the list of functions or variables inside a module
VI.	Answer: a)	To open a file
VII.	Answer: c)	Gensim
VIII.	Answer: a)	To import all functions and variables from a module
IX.	Answer: a)	To open a file and automatically close it when you are done with it
X.	Answer: c)	XGBoost

03

3 DATA PREPARATION

In machine learning, the quality and characteristics of the data plays a crucial role in the performance of the model. The data preparation step is the process of cleaning, transforming, and organizing the data to make it suitable for the machine learning model. The goal of this chapter is to provide an understanding of the data preparation process and the various techniques used to preprocess the data. We will cover data cleaning, feature scaling, feature selection, and data transformation, as well as the importance of evaluating the quality of the data. By the end of this chapter, you will have a clear understanding of how to prepare your data for machine learning, and how to ensure that it is of the highest quality.

3.1 IMPORTING DATA

Importing and cleaning data is an essential step in the machine learning process. The quality and characteristics of the data plays a crucial role in the performance of the model, and it is important to ensure that the data is in a format that can be understood by the machine learning model. In this section, we will discuss the process of importing and cleaning data, and the various techniques used to preprocess the data.

The first step in the data preparation process is to import the data into the program. In Python, there are several libraries that can be used to import data, including Pandas, NumPy, and CSV.

Pandas is a library that provides easy-to-use data structures and data analysis tools. It can be used to import data from a variety of sources, including CSV, Excel, and SQL databases.

NumPy is a library for numerical computation that provides support for large, multi-dimensional arrays and matrices. It can be used to import data in the form of arrays.

CSV (Comma Separated Values) is a common file format for storing data in a tabular form. In Python, the CSV module can be used to import data from a CSV file.

A common example of importing data in Python is using the Pandas library to import a CSV file. The following is an example of how to import a CSV file called "example_data.csv" using Pandas. The file "example_data.csv" contain employee data with following columns:

EMPLOYEE_ID, FIRST_NAME, LAST_NAME, EMAIL, PHONE_NUMBER, HIRE_DATE, JOB_ID, YEAR_OF_EXP, EDUCATION, CERTIFICATION, COMMISSION_PCT, MANAGER_ID, DEPARTMENT_ID, AGE, SALARY

The file will be available on this book GitHub repository. Here is the code for importing the csv file:

```
import pandas as pd

# Import the data from the CSV file
data = pd.read_csv("example_data.csv")
```

In this example, the first line imports the Pandas library and assigns it the alias "pd". The second line uses the "read_csv" function provided by Pandas to import the data from the "example_data.csv" file and store it in a variable called "data". The "read_csv" function automatically detects the delimiter (comma) used in the CSV file and separates the data into columns and rows.

Once the data is imported, it can be easily manipulated and analyzed. For example, you can view the first five rows of the data using the following code:

```
print(data.head())
```

You can also access specific columns and rows of the data using the following code:

```
# Access the column "AGE"
age = data["AGE"]

# Access the row with index 2
row_2 = data.loc[2]
```

In this example, we have shown how to import a CSV file into python using Pandas, and also how to access specific columns and rows of the data once imported. Pandas is a powerful library that makes it easy to import and manipulate structured data, it's easy to use and understand, it's widely adopted in data science and machine learning, and it is one of the most important libraries for data preparation.

List of function for importing data

There are several python functions that can be used to import data into a python script, including:

1. **pandas.read_csv()** - This function can be used to import data from a CSV file into a pandas DataFrame.
2. **pandas.read_excel()** - This function can be used to import data from an Excel file into a pandas DataFrame.
3. **pandas.read_json()** - This function can be used to import data from a JSON file into a pandas DataFrame.
4. **pandas.read_sql()** - This function can be used to import data from a SQL database into a pandas DataFrame.
5. **pandas.read_html()** - This function can be used to import data from an HTML file into a pandas DataFrame.
6. **pandas.read_pickle()** - This function can be used to import data from a pickle file into a pandas DataFrame.
7. **pandas.read_fwf()** - This function can be used to import data from a fixed-width-format file into a pandas DataFrame.
8. **pandas.read_stata()** - This function can be used to import data from a STATA file into a pandas DataFrame.
9. **pandas.read_sas()** - This function can be used to import data from a SAS file into a pandas DataFrame.
10. **pandas.read_clipboard()** - This function can be used to import data from the clipboard into a pandas DataFrame.

These are some of the most common functions used for importing data in python using pandas library. Depending on the format and source of the data, different functions can be used to import it into python.

In addition, the Pandas library provides many useful functions for data manipulation, such as sorting, filtering, and aggregating the data. This makes it a powerful tool for data analysis and preparation.

Once the data is imported, it can be easily manipulated and analyzed.

To explore the data importing, cleaning and transformation in more detail, you can read the book "Python for Data Analysis" by the same author. There we explain all the above process and their relevant library such as NumPy, pandas in more detail.

3.2 CLEANING DATA

After importing the data, the next step is to clean it. Data cleaning is the process of removing or correcting inaccurate, incomplete, or irrelevant data. This step is important because the quality of the data can have a significant impact on the performance of the machine learning model. The following are some common data cleaning techniques:

Removing duplicate data

Removing duplicate data is an important step in data preprocessing as it can improve the accuracy and efficiency of machine learning models. Duplicate data can occur for various reasons, such as data entry errors, data merging, or data scraping.

There are several techniques for removing duplicate data, including:

1. **Removing duplicate rows:** This method involves identifying and removing duplicate rows based on one or more columns. This method can be useful when the duplicate data is limited to a small number of rows.

2. **Removing duplicate columns:** This method involves identifying and removing duplicate columns based on one or more columns. This method can be useful when duplicate data is limited to a small number of columns.

3. **Removing duplicate records based on a subset of columns:** This method involves identifying and removing duplicate records based on a subset of columns. This method can be useful when duplicate data is limited to a specific subset of columns.

We can do this using the following code snippet in python using the pandas library:

```
# Import the data
data = pd.read_csv("example_data.csv")

# Remove duplicate rows
data = data.drop_duplicates()

# Remove duplicate columns
data = data.loc[:,~data.columns.duplicated()]

#Remove duplicate records based on a subset of columns
data = data.drop_duplicates(subset = ['EMPLOYEE_ID', 'EMAIL',
'PHONE_NUMBER'])
```

In this example, the first line imports the data from the "example_data.csv" file and store it in a variable called "data".

The second line uses the drop_duplicates() method from pandas to remove the duplicate rows from the dataframe.

The third line uses the drop_duplicates() method from pandas to remove the duplicate columns from the dataframe by using the option of **.loc[:,~data.columns.duplicated()]**

The fourth line uses the drop_duplicates() method with the subset parameter, to remove the duplicate records based on a subset of columns, in this case columns **'EMPLOYEE_ID'**, **'EMAIL'**, and **'PHONE_NUMBER'**.

It's important to keep in mind that when removing duplicate data, it's crucial to consider the size of the dataset, as removing duplicate data can cause a significant loss of information.

It's also important to check the distribution of the data after removing duplicate data to ensure that it makes sense for the data.

Sometimes it's not always necessary to remove duplicate data, for example, in case of time-series data, duplicate data can be useful for studying the trends over time. Therefore, it is important to consider the context of the data and the research question before removing duplicate data.

It's also important to keep in mind that when removing duplicate data, it's crucial to use the appropriate method based on the specific characteristics of the data and the research question. For example, if duplicate data is limited to a specific subset of columns, it's more appropriate to remove duplicate records based on that subset of columns rather than removing all duplicate data.

Handling missing data

Handling missing values is an important step in data preprocessing as it can have a significant impact on the performance of machine learning models. Missing values can occur for various reasons, such as data entry errors, measurement errors, or non-response.

There are several techniques for handling missing values, including:

1. **Deleting the rows or columns with missing values**: This is the simplest method, but it can result in a loss of important information if a large number of observations are removed.

2. **Imputing the missing values**: This method involves replacing the missing values with a substitute value, such as the mean, median, or mode of the variable. This method can be useful when the number of missing values is small, but it can introduce bias if the missing values are not missing at random.

3. **Using a predictive model to impute missing values**: This method involves training a model to predict the missing values based on the non-missing values. This method can be useful when the number of missing values is large, but it can be computationally expensive.

We can do this using the following code snippet in python using the scikit-learn library:

```
from sklearn.impute import SimpleImputer

# Import the data
data = pd.read_csv("example_data.csv")
X = (data.drop(columns =["EMPLOYEE_ID", "FIRST_NAME", "LAST_NAME", 
"EMAIL",
                         "PHONE_NUMBER", "HIRE_DATE", "JOB_ID", 
"COMMISSION_PCT",
                         "MANAGER_ID", "DEPARTMENT_ID", "SALARY"], 
axis=1))

# Create an imputer object
imputer = SimpleImputer(strategy="mean")

# Fit the imputer object to the data
imputer.fit(X)

# Transform the data
X_imputed = imputer.transform(X)
```

In this example, the first line imports the SimpleImputer class from the scikit-learn library. The second line imports the data from the "example_data.csv" file and store it in a variable called "data". The third line separates the predictor variables (X) and all non-numeric columns (namely EMPLOYEE_ID, FIRST_NAME, LAST_NAME, EMAIL, PHONE_NUMBER, HIRE_DATE, JOB_ID, , COMMISSION_PCT, MANAGER_ID, DEPARTMENT_ID, SALARY) from the target variable. The fourth line creates an imputer object and sets the strategy parameter to "mean" which means that the missing values will be replaced with the mean value of the feature. The fifth line fits the imputer object to the data. The sixth line uses the transform method to replace the missing values with the mean value of the feature.

It's important to keep in mind that the chosen method for handling missing values should be based on the specific characteristics of the data and the research question. Additionally, it's also important to check the distribution of the data after handling missing values to ensure that it makes sense for the data.

Handling outliers

Handling outliers is an important step in data preprocessing as it can have a significant impact on the performance of machine learning models. Outliers are observations that deviate significantly from the majority of the data. They can occur for various reasons, such as measurement errors, data entry errors, or data from a different distribution.

There are several techniques for handling outliers, including:

1. **Removing outliers**: This method involves identifying and removing observations that deviate significantly from the majority of the data. This

method can be useful when the number of outliers is small, but it can result in a loss of important information if a large number of observations are removed.

2. **Transforming the data**: This method involves transforming the data using techniques such as log transformation, square root transformation, or reciprocal transformation to reduce the impact of outliers.
3. **Imputing outliers**: This method involves replacing outliers with a substitute value, such as the mean, median, or mode of the variable. This method can be useful when the number of outliers is small, but it can introduce bias if the outliers are not missing at random.
4. **Using robust models**: This method involves using models that are less sensitive to outliers such as decision trees or linear discriminant analysis.

We can do this using the following code snippet in python using the scikit-learn library:

```
from sklearn.covariance import EllipticEnvelope

# Import the data
data = pd.read_csv("example_data.csv")
X = (data.drop(columns =["EMPLOYEE_ID", "FIRST_NAME", "LAST_NAME",
                         "EMAIL",
                         "PHONE_NUMBER", "HIRE_DATE", "JOB_ID",
                         "COMMISSION_PCT",
                         "MANAGER_ID", "DEPARTMENT_ID", "SALARY"],
                         axis=1))

########### Handling missing values ###############
# Create an imputer object
imputer = SimpleImputer(strategy="mean")

# Fit the imputer object to the data
imputer.fit(X)

# Transform the data
```

```
X_imputed = imputer.transform(X)
X = X_imputed

########### Handling outliers #####################
# Create an EllipticEnvelope object
outlier_detector = EllipticEnvelope(contamination=0.05)

# Fit the outlier detector to the data
outlier_detector.fit(X)

# Predict the outliers
y_pred = outlier_detector.predict(X)

# Identify the outliers
outliers = X[y_pred == -1]
```

In this example, the first line imports the EllipticEnvelope class from the scikit-learn library. The second line imports the data from the "example_data.csv" file and store it in a variable called "data". The third line separates the predictor variables (X) and non-numeric variable from the target variable. After that, we handle missing values as described in previous section.

The next line after comment (*########### Handling outliers #####################*) creates an outlier detector object and sets the contamination parameter to 0.05 which means that 5% of the data is considered as outliers. The next line fits the outlier detector to the data. The next line uses the predict method to identify the observations that deviate significantly from the majority of the data. The next line identifies the outliers by using the predictions from the outlier detector.

It's important to keep in mind that the chosen method for handling outliers should be based on the specific characteristics of the data and the research question. Additionally, it's also important to check the distribution of the data after handling outliers to ensure that it makes sense for the data.

Formatting data

Formatting data is an important step in data preprocessing as it ensures that the data is in a consistent and usable format for machine learning models. Formatting data involves converting data into a format that can be easily consumed by machine learning algorithms. This can include tasks such as converting data types, encoding categorical variables, and standardizing variable names.

There are several techniques for formatting data, including:

1. Converting data types: This method involves converting data into the appropriate data type, such as converting text data into numerical data or converting date/time data into a timestamp format. This can be done using functions such as **to_numeric**, **to_datetime** in pandas library.

2. Encoding categorical variables: This method involves converting categorical variables, such as text data, into a numerical format that can be used by machine learning models. This can be done using techniques such as one-hot encoding, ordinal encoding, or dummy encoding.

3. Standardizing variable names: This method involves converting variable names into a consistent format, such as converting variable names to lowercase or removing spaces. This can be done using functions such as **str.lower()**, **str.strip()** in pandas library.

We can do this using the following code snippet in python using the pandas library:

```
# Import the data
data = pd.read_csv("example_data.csv")

#Convert data types
```

```
data['AGE'] = data[['AGE']].astype(int)
data['HIRE_DATE'] = pd.to_datetime(data['HIRE_DATE'])

#Encoding categorical variables
data = pd.get_dummies(data, columns=["EDUCATION_LEVEL"])

#Standardizing variable names
data.rename(columns={'EMAIL': 'EMAIL_ID'}, inplace=True)
data.columns = data.columns.str.strip().str.lower().str.replace(' ',
'_')
```

In this example, the first line imports the data from the "example_data.csv" file and store it in a variable called "data". The second line converts the data types of column 'age' to integer using the **astype()** method and the column 'date' to datetime using the **pd.to_datetime()** method. The third line uses the **pd.get_dummies()** method to encode the categorical variable 'color' by one-hot encoding The fourth line uses the **rename()** method to standardize the variable name 'name' to 'full_name' and also use **str.strip()**, **str.lower()**, **str.replace()** to standardize all columns name to lowercase, remove spaces and replace spaces with underscore.

It's important to keep in mind that the chosen method for formatting data should be based on the specific characteristics of the data and the research question. Additionally, it's also important to check the distribution of the data after formatting data to ensure that it makes sense for the data.

In conclusion, Formatting data is an important step in data preprocessing as it ensures that the data is in a consistent and usable format for machine learning models. Formatting data involves converting data into a format that can be easily consumed by machine learning algorithms. This can include tasks such as converting data types, encoding categorical variables, and standardizing variable names. There are several techniques for formatting data, including converting data types, encoding categorical variables, and standardizing variable names. The chosen method should be based on the specific characteristics of the data and the research question. Formatting data correctly can help to improve the performance of machine learning models and make the data easier to work with.

It's also important to check the data for any inconsistencies or errors, such as typos or mislabeled data, and to correct them as necessary.

It's important to note that data cleaning can be a time-consuming process, but it is critical to improve the performance of the machine learning model.

List of function to clean data

There are several python functions that can be used to clean data in a python script, including:

1. **pandas.DataFrame.drop()** - This function can be used to drop specified labels from rows or columns. It can be used to drop rows or columns based on their index or column name, and can also be used to drop rows or columns based on certain conditions.

2. **pandas.DataFrame.fillna()** - This function can be used to fill missing values with a specific value or method. It can be used to fill missing values with a specific value, such as the mean or median of the data, or it can be used to forward-fill or backward-fill missing values.

3. **pandas.DataFrame.replace()** - This function can be used to replace values in a DataFrame. It can be used to replace a specific value or a set of values with another value or set of values.

4. **pandas.DataFrame.drop_duplicates()** - This function can be used to remove duplicate rows from a DataFrame. It can be used to drop duplicate

rows based on one or more columns and can also be used to keep the first or last occurrence of duplicate rows.

5. **pandas.DataFrame.query()** - This function can be used to filter rows of a DataFrame based on a Boolean expression. It can be used to filter rows based on certain conditions, such as removing rows with missing values or removing rows with specific values.

6. **pandas.DataFrame.rename()** - This function can be used to rename columns or row indexes of a DataFrame. It can be used to rename one or multiple columns or indexes.

7. **pandas.DataFrame.sort_values()** - This function can be used to sort a DataFrame by one or more columns. It can be used to sort the data in ascending or descending order and can also be used to sort based on multiple columns.

8. **pandas.DataFrame.groupby()** - This function can be used to group rows of a DataFrame based on one or more columns. It can be used to group data by a specific column and perform calculations such as mean, sum, or count on the grouped data.

For example,

```
import pandas as pd

# Import the data
data = pd.read_csv("example_data.csv")

# Remove missing values
data = data.dropna()

# Replace specific values
data = data.replace({'JOB_ID': {'ST_CLERK': 'STATION_CLERK'}})
```

```
# Remove duplicate rows
data = data.drop_duplicates()

# Rename columns
data = data.rename(columns={'FIRST_NAME': 'GIVEN_NAME'})
```

In this example, the first line imports the data from the "example_data.csv" file and store it in a variable called "data". The second line uses the **dropna()** function to remove all the rows that contain missing values. The third line uses the **replace()** function to replace all occurrences of the value 'green' in the 'color' column with 'blue'. The fourth line uses the **drop_duplicates()** function to remove all duplicate rows from the DataFrame. The fifth line uses the **rename()** function to rename the column 'name' to 'full_name'.

It's important to keep in mind that the chosen method for cleaning data should be based on the specific characteristics of the data and the research question. Additionally, it's also important to check the distribution of the data after cleaning to ensure that it makes sense for the data. It's also important to keep in mind that data cleaning is an iterative process and it's necessary to repeat the process until the data is in the desired format.

A common example of cleaning data in Python is using the Pandas library to handle missing values. The following is an example of how to handle missing values in a DataFrame:

```
import pandas as pd

# Import the data from the CSV file
data = pd.read_csv("example_data.csv")

# Checking the number of missing values in each column
print(data.isnull().sum())
```

```
# Dropping rows with missing values
data = data.dropna()

# Filling missing values with mean of the column
data = data.fillna(data.mean())
```

In this example, the first line imports the Pandas library and assigns it the alias "pd". The second line uses the "read_csv" function provided by Pandas to import the data from the "example_data.csv" file and store it in a variable called "data".

The third line uses the "isnull()" function provided by Pandas to check for missing values in each column of the data, and the "sum()" function to count the number of missing values in each column.

The fourth line uses the "dropna()" function to drop all the rows that contain missing values. This method is useful when the missing values are in a small number and can be dropped without affecting the overall data.

The fifth line uses the "fillna()" function to fill missing values with the mean of the column. This method is useful when the missing values are in a large number and need to be replaced with a suitable value, in this case, the mean of the column.

It's important to note that there are many other ways to handle missing values such as using median, mode, or interpolation methods. The choice of method depends on the characteristics of the data and the specific requirements of the project.

In this example, we have shown how to handle missing values in a DataFrame using the Pandas library. We have used the "isnull()" and "sum()" functions to check for missing values, the "dropna()" function to drop rows with missing values, and the "fillna()" function to fill missing values with the mean of the column. These are just some of the ways to handle missing values, and different methods may work better depending on the characteristics of the data.

It's important to keep in mind that data cleaning is an iterative process and it may require multiple steps to clean the data properly. It's also important to check the data for any inconsistencies or errors, such as typos or mislabeled data, and to correct them as necessary.

In conclusion, cleaning data is an essential step in the machine learning process, it is important to ensure that the data is of high quality. This can be achieved by using various libraries and techniques such as Pandas, handling missing values, dropping rows, filling missing values with suitable values, checking for inconsistencies and errors, these steps are crucial to improve the performance of the machine learning model. The more time and effort you invest in data cleaning, the better the performance of your machine learning model will be.

3.3 EXPLORATORY DATA ANALYSIS

Exploratory Data Analysis (EDA) is a technique used to understand and analyze data. The goal of EDA is to gain insights and understand the underlying structure of the data, as well as identify any patterns, relationships, or anomalies that may be present. EDA is an iterative process, and it is typically the first step in the data analysis pipeline. In this section, we will discuss the process of EDA, and the various techniques used to explore and understand the data.

Univariate Analysis

The first step in EDA is to perform univariate analysis, which involves **analyzing each variable individually**. This helps to understand the distribution of the data and identify any outliers or anomalies. Some common techniques used in univariate analysis include:

Descriptive statistics

Descriptive statistics are a crucial part of data analysis that allows us to summarize and describe the main characteristics of a dataset. In univariate analysis, descriptive statistics are used to summarize and describe the properties of a single variable.

Descriptive statistics in univariate analysis can be classified into two broad categories: measures of central tendency and measures of variability.

Measures of Central Tendency

Measures of central tendency are used to describe the typical or central value of a distribution. The most common measures of central tendency are:

1. **Mean**: It is the sum of all observations in the dataset divided by the total number of observations.
2. **Median**: It is the middle value in a dataset. When a dataset has an even number of observations, the median is the average of the two middle values.
3. **Mode**: It is the value that occurs most frequently in a dataset.

Measures of Variability

Measures of variability are used to describe the spread or dispersion of the data. The most common measures of variability are:

1. **Range**: It is the difference between the maximum and minimum values in a dataset.

2. **Variance**: It is the average of the squared differences of each value from the mean.

3. **Standard Deviation**: It is the square root of the variance and is used to describe the spread of data around the mean.

Other measures of variability include percentiles, which are used to describe the distribution of the data over the entire range.

Descriptive statistics can be presented using different types of graphical representations such as histograms, boxplots, and scatterplots. These visualizations help to identify patterns and outliers in the data.

A common example of univariate analysis is using the Pandas library to calculate descriptive statistics of a dataset. The following is an example of how to perform univariate analysis on a variable "Age" in a dataset "data":

```
import pandas as pd

# Import the data
data = pd.read_csv("example_data.csv")

# Extract the variable "Age"
age = data["AGE"]

# Calculate descriptive statistics
print("Mean: ", age.mean())
print("Median: ", age.median())
print("Mode: ", age.mode())
print("Standard Deviation: ", age.std())
print("Minimum Value: ", age.min())
print("Maximum Value: ", age.max())
```

In this example, the first line imports the Pandas library and assigns it the alias "pd". The second line uses the "read_csv" function provided by Pandas to import the data from the "example_data.csv" file and store it in a variable called "data".

The third line uses the bracket notation to extract the variable "AGE" from the dataframe and store it in a variable called "age".

The fourth line uses the mean() function to calculate the mean of the variable "Age", the median() function to calculate the median of the variable, the mode() function to calculate the mode of the variable, the std() function to calculate the standard deviation of the variable, the min() function to calculate the minimum value of the variable, and the max() function to calculate the maximum value of the variable.

The results of the descriptive statistics can be used to understand the distribution of the variable "Age", for example, if the mean is close to the median, it indicates that the variable is distributed normally, if the mean and median are far apart it indicates that the variable is distributed skew. The standard deviation can also be used to understand the variability of the variable, where a low standard deviation indicates that the variable is distributed closely around the mean, while a high standard deviation indicates that the variable is distributed widely around the mean.

In this example, we have shown how to perform univariate analysis using the Pandas library, specifically calculating descriptive statistics of a variable. This is just one of the many techniques that can be used to perform univariate analysis, and different methods may work better depending on the characteristics of the data.

In conclusion, descriptive statistics in univariate analysis are essential for summarizing and describing the properties of a single variable. These statistics provide valuable insights into the distribution of the data and help to identify any patterns or outliers. Data analysts and data scientists use these measures to make informed decisions about the data and to build models for data prediction and forecasting.

Histograms

Histograms are a graphical representation of the distribution of numerical data. They display the frequency or proportion of values that fall within specific ranges or bins. In univariate analysis, histograms are used to visualize the distribution of a single variable.

For example, let's say we have a dataset that contains the heights of a group of people. We can create a histogram to visualize the distribution of these heights. The histogram will show the frequency or proportion of people with heights in each bin.

To create a histogram in Python, we can use the Matplotlib library. Here's an example code snippet:

```
import matplotlib.pyplot as plt
import numpy as np

# Generate some random data
data = np.random.normal(size=1000)

# Create histogram
plt.hist(data, bins=30, alpha=0.5, color='b')

# Add labels and title
plt.xlabel('Height')
plt.ylabel('Frequency')
plt.title('Distribution of Heights')

# Show plot
plt.show()
```

The output look like this:

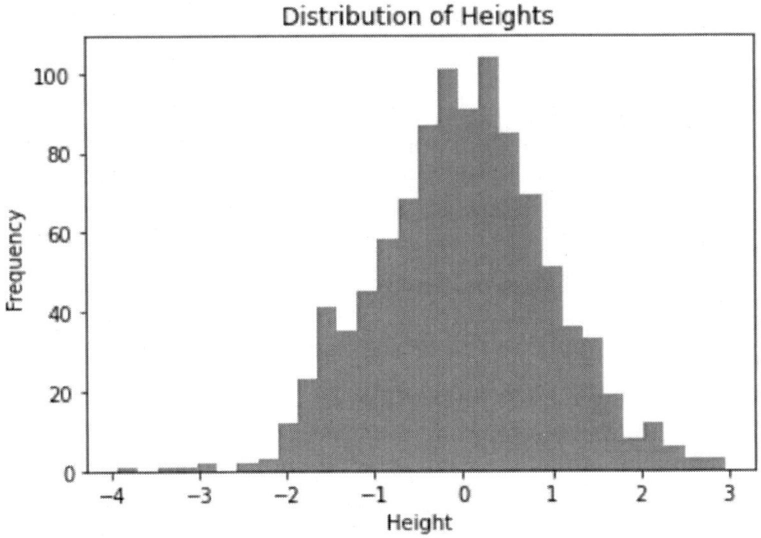

In this example, we first generate some random data using the NumPy library. We then create a histogram using the **plt.hist()** function, specifying the number of bins to use, the transparency and color of the bars, and other properties. We then add labels and a title to the plot and display it using **plt.show()**.

The resulting histogram will display the distribution of the heights in the data, with the x-axis showing the height range and the y-axis showing the frequency or proportion of people with heights in that range.

Histograms are a useful tool in univariate analysis as they allow us to quickly visualize the distribution of a variable and identify patterns or outliers in the data.

Box plots

In statistics, univariate analysis involves the examination of a single variable. One way to visually summarize the distribution of a single variable is through box plots. A box plot, also known as a box and whisker plot, displays a summary of the data's median, quartiles, and range. This graph is useful in identifying outliers and comparing the distribution of different datasets.

To construct a box plot, you need the minimum value, lower quartile (Q1), median, upper quartile (Q3), and maximum value. The interquartile range (IQR) is the distance between the upper and lower quartiles. The box represents the IQR, with the median shown as a line inside the box. The whiskers extend from the box to the minimum and maximum values, excluding outliers.

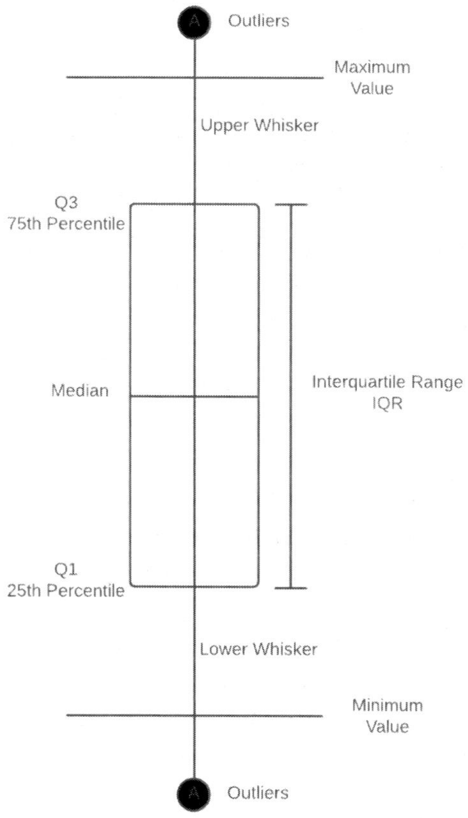

Here is an example of a box plot for a dataset of exam scores:

```
Dataset: 40, 60, 70, 75, 80, 85, 90, 95, 100

Minimum value: 40
Lower quartile (Q1): 70
Median: 80
Upper quartile (Q3): 90
Maximum value: 100
IQR: Q3 - Q1 = 20

  40    60    70    75    80    85    90    95   100
  |     |_____|           |___|
       Q1    Median       Q3

Whiskers: 40 and 100
Outliers: None
```

In this example, the dataset has a minimum value of 40, a lower quartile of 70, a median of 80, an upper quartile of 90, and a maximum value of 100. The IQR is calculated as 20. The box plot shows that the majority of the data falls between 70 and 90, with two scores outside this range (75 and 100).

Box plots can also be used to compare multiple datasets. In the case of multiple datasets, each box plot is drawn side by side, making it easy to visually compare their median, IQR, and range. This comparison can help to identify differences in the distribution of the data.

here is an example of how to create a box plot in Python using the Matplotlib library:

```
import matplotlib.pyplot as plt
import numpy as np

# generate some random data
data = np.random.normal(size=100)

# create a box plot of the data
```

```
fig, ax = plt.subplots()
ax.boxplot(data)

# set the title and labels
ax.set_title('Box Plot of Random Data')
ax.set_ylabel('Values')

# show the plot
plt.show()
```

The output will look like this

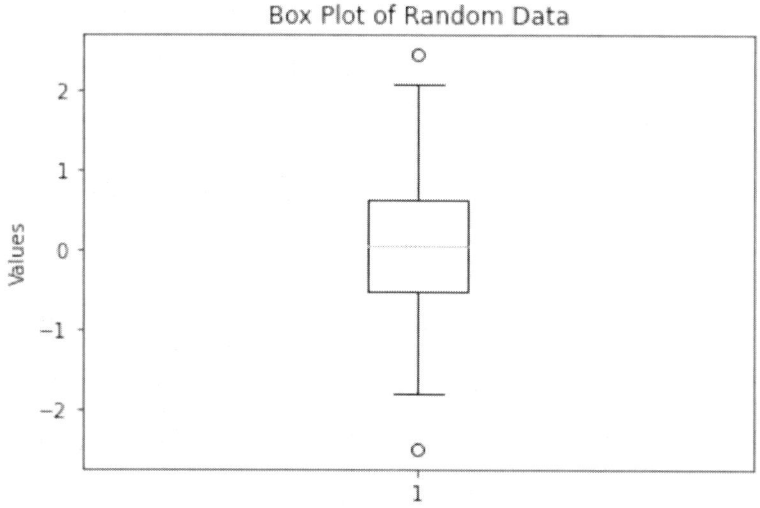

In this example, we first import the necessary libraries: **matplotlib** and **numpy**. We then generate some random data using the **numpy.random.normal()** function. This generates an array of 100 values sampled from a normal distribution with a mean of 0 and standard deviation of 1.

Next, we create a figure and axis object using the **subplots()** function. We then call the **boxplot()** function on the axis object, passing in the data as a parameter. This creates a box plot of the data.

We then set the title and y-axis label using the **set_title()** and **set_ylabel()** methods on the axis object, respectively.

Finally, we call the **show()** function to display the plot.

The resulting plot should show a box with a line in the middle (the median), a box on either side of the median representing the interquartile range (IQR), and whiskers extending from the boxes to show the minimum and maximum values that are within 1.5 times the IQR from the box. Any values beyond the whiskers are considered outliers and are plotted as individual points.

In summary, box plots are a useful tool for visualizing the distribution of a single variable or comparing the distributions of multiple variables. They provide a concise summary of the data's median, quartiles, and range, as well as identifying potential outliers.

Bivariate Analysis

Once the univariate analysis is complete, the next step is to perform bivariate analysis, which involves **analyzing the relationship between two variables**. This helps to understand how the variables are related and identify any patterns or correlations in the data. Some common techniques used in bivariate analysis include:

Scatter plots

In statistical analysis, bivariate analysis refers to the analysis of two variables. One common method of visualizing bivariate data is through a scatter plot. A scatter plot is a graph in which the values of two variables are plotted along two axes, with each individual data point represented by a dot on the graph. The position of the dot on the graph indicates the value of the two variables for that particular data point.

For example, suppose we want to analyze the relationship between the height and weight of a group of individuals. We can create a scatter plot with height on the x-axis and weight on the y-axis, with each individual represented by a dot on the graph.

In the resulting scatter plot, each dot represents a single individual, and the position of the dot on the graph indicates both the height and weight of that individual. By looking at the scatter plot, we can see the overall pattern of the relationship between the two variables.

If the dots on the graph form a roughly linear pattern, we can say that there is a positive correlation between the two variables, which means that as one variable increases, so does the other. On the other hand, if the dots on the graph are spread out randomly with no discernible pattern, we can say that there is no correlation between the two variables.

In addition to examining the overall pattern of the relationship between the two variables, scatter plots can also be used to identify outliers, which are individual data points that fall far outside the range of the other data points.

Here's an example code for creating a scatter plot using the **matplotlib** library in Python:

```
import matplotlib.pyplot as plt

# Sample data
x = [1, 2, 3, 4, 5, 6, 7, 8, 9, 9, 8, 7, 6, 5, 4, 3, 3]
y = [3, 5, 2, 7, 4, 2, 4, 5, 5, 6, 2, 3, 4, 5, 7, 2, 1]

# Create scatter plot
plt.scatter(x, y)

# Add axis labels and title
plt.xlabel('X-axis')
plt.ylabel('Y-axis')
```

```
plt.title('Scatter Plot Example')

# Show the plot
plt.show()
```

In this example, we first import the **matplotlib.pyplot** module and create two lists of data, **x** and **y**. We then call the **scatter()** function from **matplotlib.pyplot** to create the scatter plot using the data.

Next, we add axis labels and a title to the plot using the **xlabel()**, **ylabel()**, and **title()** functions.

Finally, we call the **show()** function to display the plot.

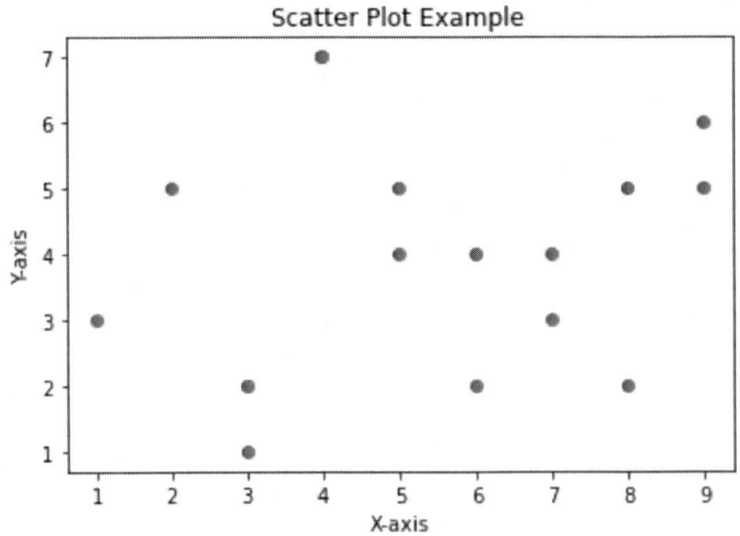

The resulting scatter plot will have the **x** values on the horizontal axis and the **y** values on the vertical axis, with a point representing each pair of values. The axis labels and title provide additional information about the data being plotted.

Overall, scatter plots provide a useful way to explore the relationship between two variables and identify any patterns or outliers in the data.

Correlation

In statistics, correlation is a measure of the strength and direction of the relationship between two variables. Correlation analysis is used in bivariate analysis to identify the extent to which two variables are related to each other.

For example, suppose we are interested in examining the relationship between the age of a car and its price. We can use correlation analysis to determine if there is a relationship between these two variables, and if so, whether it is a positive or negative relationship.

Correlation is usually measured using a correlation coefficient, which ranges from -1 to +1. A coefficient of +1 indicates a perfect positive correlation, a coefficient of -1 indicates a perfect negative correlation, and a coefficient of 0 indicates no correlation.

There are two main types of correlation: Pearson correlation and Spearman correlation. Pearson correlation is used when both variables are normally distributed, while Spearman correlation is used when one or both variables are not normally distributed.

Here is an example of using Pearson correlation to examine the relationship between the age of a car and its price in Python:

```
import numpy as np
import pandas as pd
import matplotlib.pyplot as plt

# create a sample dataset
age = np.array([1, 3, 5, 7, 9])
price = np.array([10000, 9000, 7000, 5000, 3000])
```

```
# calculate the correlation coefficient
corr_coef = np.corrcoef(age, price)[0, 1]

# plot the data and the regression line
plt.scatter(age, price)
plt.plot(age, np.poly1d(np.polyfit(age, price, 1))(age))
plt.xlabel('Age (years)')
plt.ylabel('Price ($)')
plt.title(f'Correlation Coefficient: {corr_coef:.2f}')
plt.show()
```

In this example, we first create a sample dataset with the age of the car in years and its price in dollars. We then calculate the Pearson correlation coefficient using the **np.corrcoef** function, and plot the data using **plt.scatter**. Finally, we plot the regression line using **np.polyfit** and **plt.plot**, and display the correlation coefficient in the title using string formatting.

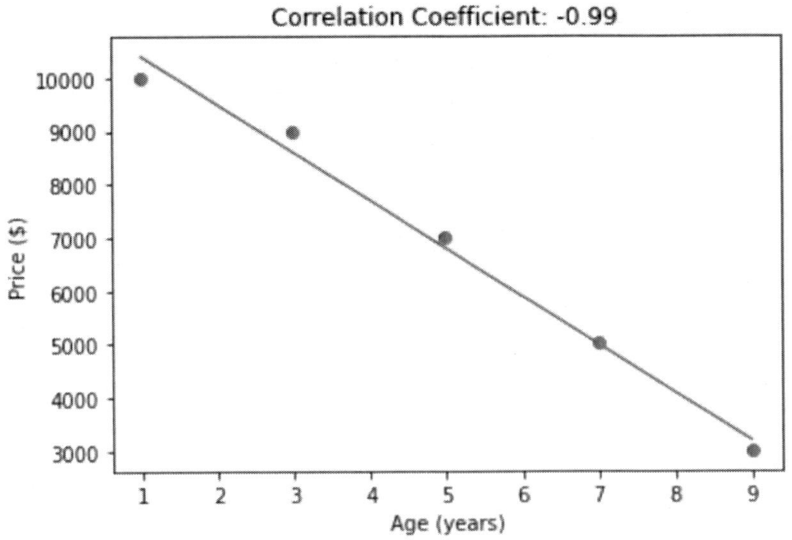

The resulting plot shows that there is a strong negative correlation between the age of the car and its price, with a correlation coefficient of -0.99. This indicates that as the age of the car increases, its price decreases.

This is just one of the many techniques that can be used to perform bivariate analysis, and different methods such as heatmap, line plot, bar plot, etc. may work better depending on the characteristics of the data. It's important to note that bivariate analysis is a powerful tool that can be used to identify patterns, relationships, and correlations between variables, which can be useful for understanding the underlying structure of the data and guiding further analysis.

Multivariate Analysis

After performing univariate and bivariate analysis, the next step is to perform multivariate analysis, which involves analyzing the relationship among three or more variables. This helps to understand how the variables are related and identify any patterns or correlations in the data. Some common techniques used in multivariate analysis include:

Heatmaps

A heatmap is a graphical representation of data that uses a color-coding system to represent different values. It is a useful tool for visualizing the relationships between multiple variables in a dataset. In multivariate analysis, a heatmap is used to display the correlation between multiple variables, making it easier to identify patterns and trends in large datasets.

To create a heatmap in Python, we can use the **seaborn** library. First, we need to generate some random data to work with. We can use the **numpy** library to generate random data and set a seed to ensure that the data is the same each time the code is run.

Here is an example of how a heatmap can be used in multivariate analysis:

Suppose we want to understand the relationship between student performance and various factors such as study time, family income, and parental education level. We can create a heatmap to visualize the correlation between these variables.

To generate the data, we can use the NumPy library and set a seed to ensure the data is consistent each time. Here is an example code:

```
import numpy as np
import matplotlib.pyplot as plt

np.random.seed(42)

# generate random data
study_time = np.random.normal(6, 2, 100)
family_income = np.random.normal(50000, 15000, 100)
parent_education = np.random.normal(12, 3, 100)
test_score = study_time * 10 + family_income * 0.001 +
parent_education * 3 + np.random.normal(0, 5, 100)

# plot heatmap
plt.figure(figsize=(8, 6))
correlation_matrix = np.corrcoef([study_time, family_income,
parent_education, test_score])
heatmap = plt.imshow(correlation_matrix, cmap='coolwarm',
interpolation='nearest')
plt.colorbar(heatmap)
plt.xticks([0, 1, 2, 3], ['Study Time', 'Family Income', 'Parent
Education', 'Test Score'])
plt.yticks([0, 1, 2, 3], ['Study Time', 'Family Income', 'Parent
Education', 'Test Score'])
plt.title('Correlation Matrix Heatmap')
plt.show()
```

In this example, we generate 100 random values for each of the variables using the **np.random.normal()** function, which generates values from a normal distribution. We then calculate the test score based on a linear combination of these variables plus some random noise. Finally, we use the **np.corrcoef()** function to calculate the correlation coefficients between the variables, which we plot using the **plt.imshow()** function.

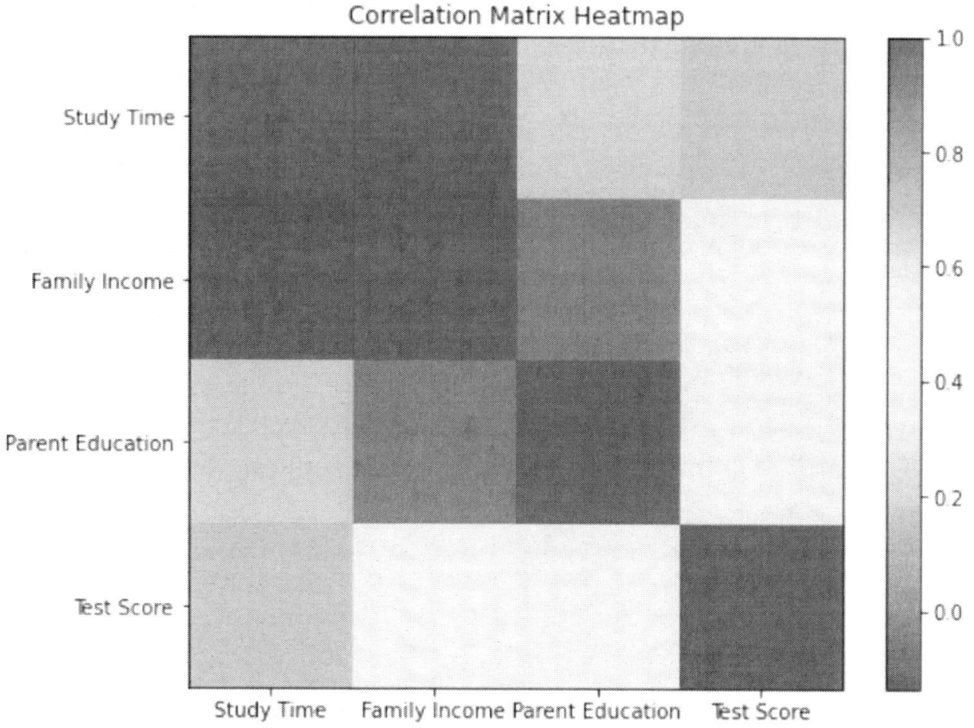

The resulting heatmap will show the correlation between each pair of variables. The darker the color, the higher the correlation. For example, we can see that there is a strong positive correlation between study time and test score, and a weaker positive correlation between family income and test score. There is also a weak negative correlation between parent education and test score. Overall, the heatmap gives us a quick and easy way to visualize the relationship between multiple variables.

Parallel coordinates

Parallel coordinates is a powerful technique for visualizing high-dimensional data. In this technique, each variable is represented by a vertical axis, and a line is drawn connecting the values of each variable for each data point. This allows us to identify patterns and relationships between variables that may not be apparent in other types of visualizations.

Here's an example scenario where parallel coordinates can be used in multivariate analysis:

Suppose we have a dataset of customer reviews for a restaurant, with features such as food quality, service, ambiance, price, and overall rating. We want to explore the relationship between these features and the overall rating given by the customers.

We can create a parallel coordinates plot to visualize this relationship. Here's an example code using Python's matplotlib library:

```
import numpy as np
import matplotlib.pyplot as plt

np.random.seed(42)

# Generate random data for 5 features and 100 samples
food_quality = np.random.randint(1, 6, size=100)
service = np.random.randint(1, 6, size=100)
ambiance = np.random.randint(1, 6, size=100)
price = np.random.randint(1, 6, size=100)
overall_rating = np.random.randint(1, 6, size=100)

# Create parallel coordinates plot
fig, ax = plt.subplots()
plt.title('Customer Reviews for a Restaurant')
plt.xlabel('Features')
plt.ylabel('Rating')
plt.xticks(range(5), ['Food Quality', 'Service', 'Ambiance', 'Price',
'Overall Rating'])
```

```
plt.ylim(0, 5)
plt.plot([0, 1, 2, 3, 4], [food_quality, service, ambiance, price,
overall_rating], color='blue', alpha=0.1)
plt.plot([0, 1, 2, 3, 4], [np.mean(food_quality), np.mean(service),
np.mean(ambiance), np.mean(price), np.mean(overall_rating)],
color='red', alpha=0.9)

plt.show()
```

In this example, we first generated random data for the five features and overall rating for 100 customer reviews, using the **numpy.random.randint()** function with a seed of 42 to ensure the same data is generated each time.

We then created a parallel coordinates plot using matplotlib's **plot()** function, where each feature is plotted as a separate line and the overall rating is shown on the y-axis. The blue lines represent individual customer reviews, while the red line represents the mean rating for each feature.

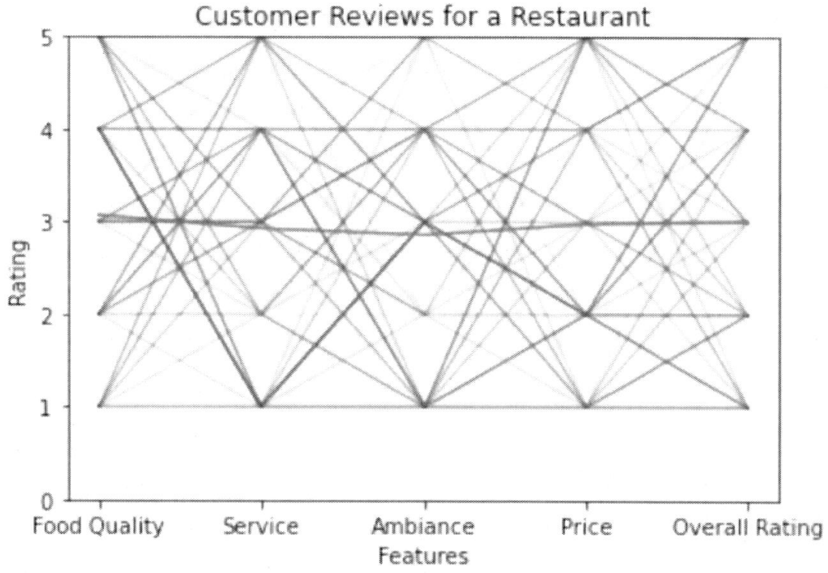

From this plot, we can observe the following:

- Customers tend to rate food quality and service higher than ambiance and price.
- The overall rating is positively correlated with food quality and service, but less so with ambiance and price.
- There is a wide range of ratings for each feature, indicating that different customers have different preferences and priorities.

This type of visualization can be useful for quickly identifying patterns and relationships in multivariate data, and can help guide further analysis and decision-making.

Cluster Analysis

Cluster analysis, also known as clustering, is a technique used in multivariate analysis to group a set of objects in a way that objects in the same group (called a cluster) are more similar to each other than to those in other groups. The objective of cluster analysis is to find a structure or pattern in the data that can be useful in discovering relationships or identifying groups within the data.

Cluster analysis can be used for a wide range of applications in different fields, such as market segmentation, image segmentation, anomaly detection, customer profiling, biological classification, and many others.

There are different methods of cluster analysis, but the most common ones are hierarchical clustering and k-means clustering.

Hierarchical Clustering

Hierarchical clustering is a method of cluster analysis that builds a hierarchy of clusters by recursively dividing the data set into smaller and smaller groups.

There are two types of hierarchical clustering: agglomerative and divisive. Agglomerative clustering starts with each data point as its own cluster and then successively merges the closest pairs of clusters until all the data points belong to a single cluster. Divisive clustering starts with all the data points in a single cluster and then successively divides the cluster into smaller clusters until each data point belongs to its own cluster.

Here's an example of hierarchical clustering using Python's **scipy** library:

```
import numpy as np
from scipy.cluster import hierarchy
import matplotlib.pyplot as plt

# Generate some random data
np.random.seed(123)
x = np.random.rand(10, 2)

# Perform hierarchical clustering
dendrogram = hierarchy.dendrogram(hierarchy.linkage(x, method='ward'))

# Visualize the dendrogram
plt.title('Hierarchical Clustering Dendrogram')
plt.xlabel('Data point')
plt.ylabel('Distance')
plt.show()
```

In this example, we first generate a random dataset with 10 data points and 2 features. We then use the **hierarchy.linkage** function from **scipy** to perform hierarchical clustering using the "ward" method. Finally, we use the **hierarchy.dendrogram** function to generate a visualization of the dendrogram.

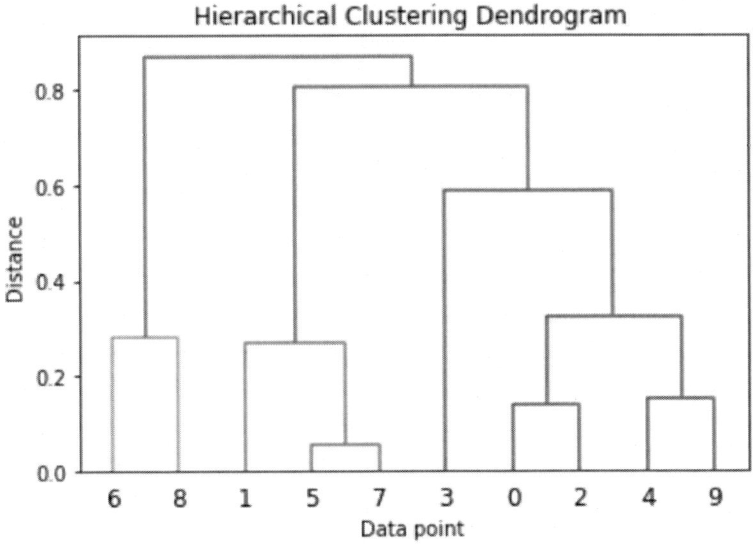

The resulting plot shows the hierarchical clustering dendrogram, where the y-axis represents the distance between the clusters being merged and the x-axis represents the data points being clustered. The height of each horizontal line in the dendrogram indicates the distance between the two clusters being merged. We can use this plot to determine the optimal number of clusters to use in our analysis, by selecting a horizontal line that cuts through the longest vertical line without intersecting any other lines.

K-means Clustering

K-means clustering is a method of cluster analysis that partitions the data set into k clusters by minimizing the sum of squared distances between each data point and the centroid of its cluster. The algorithm starts by randomly assigning each data point to one of the k clusters, and then iteratively updates the centroids and re-assigns the data points to the closest cluster until convergence.

Here's an example of how to perform cluster analysis using K-Means algorithm in Python:

```
# Import libraries
import numpy as np
import pandas as pd
import matplotlib.pyplot as plt
from sklearn.cluster import KMeans

# Generate some random data
np.random.seed(123)
x = np.random.normal(0, 1, 100)
y = np.random.normal(0, 1, 100)
z = np.random.normal(0, 1, 100)
data = pd.DataFrame({'X': x, 'Y': y, 'Z': z})

# Perform cluster analysis
kmeans = KMeans(n_clusters=3, random_state=0).fit(data)
labels = kmeans.labels_
centroids = kmeans.cluster_centers_

# Plot the clusters
colors = ['r', 'g', 'b']
fig = plt.figure(figsize=(10, 10))
ax = fig.add_subplot(111, projection='3d')
for i in range(len(data)):
    ax.scatter(data['X'][i], data['Y'][i], data['Z'][i],
c=colors[labels[i]], alpha=0.8)
ax.scatter(centroids[:, 0], centroids[:, 1], centroids[:, 2],
marker='*', s=300, c='#050505')
ax.set_xlabel('X')
ax.set_ylabel('Y')
ax.set_zlabel('Z')
plt.show()
```

In this example, we first import the necessary libraries including **numpy**, **pandas**, **matplotlib**, and **sklearn.cluster** which contains the KMeans algorithm.

We then generate random data using **numpy.random.normal** function and create a Pandas dataframe. We then perform cluster analysis on this data using the KMeans algorithm with **n_clusters=3** (i.e. we want to divide the data into 3 clusters) and **random_state=0** (to make the results reproducible).

Finally, we plot the clusters using **matplotlib** where each cluster is shown in a different color and the centroids of the clusters are shown as stars. The plot is in 3D to show the clusters in a multi-dimensional space.

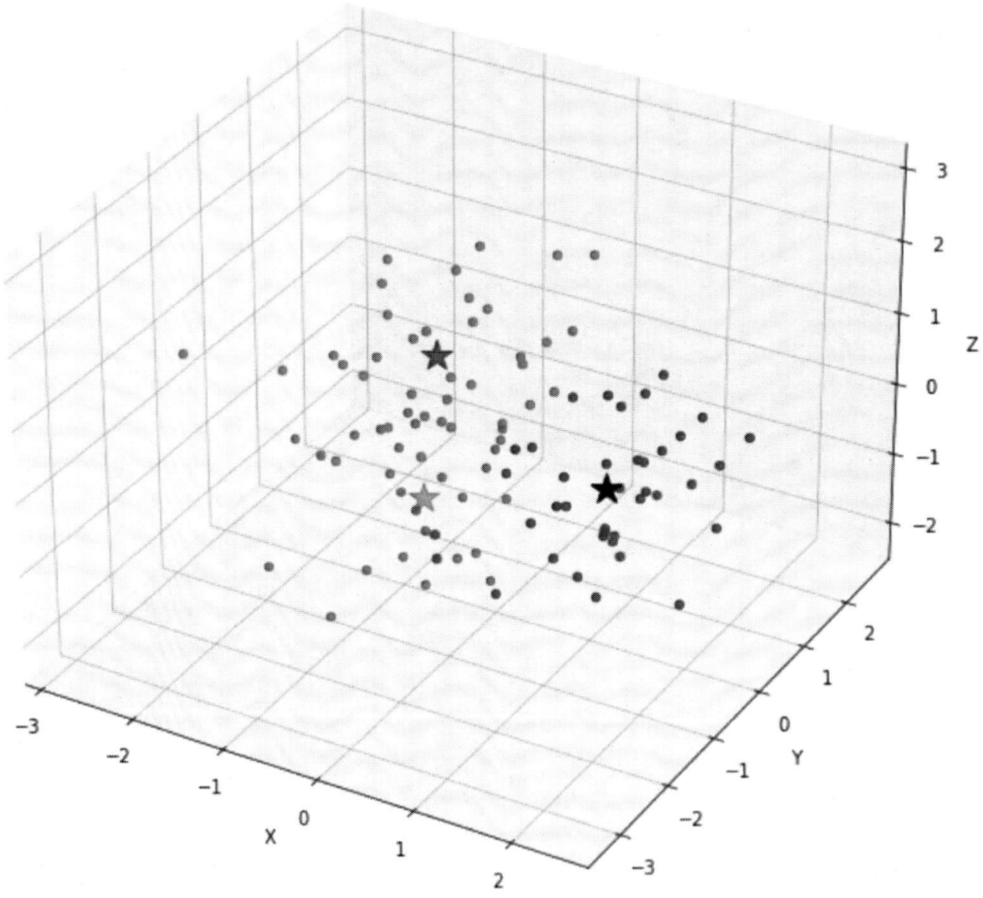

This is just a simple example, but cluster analysis can be used in various fields like customer segmentation, fraud detection, image segmentation, and more.

Both hierarchical clustering and k-means clustering have their strengths and weaknesses, and the choice of method depends on the specific problem and data set.

For example, let's say we have a data set of customers who purchased different products from an online store. We want to segment the customers into different groups based on their purchase behavior. We can use cluster analysis to identify groups of customers who are similar in their purchase behavior, and then use this information to tailor marketing campaigns and product offerings to each group.

In this case, we can use k-means clustering to partition the customers into k clusters based on their purchase behavior, and then use the cluster centroids to describe the purchase behavior of each group. We can also use hierarchical clustering to build a hierarchy of clusters, which can be useful for exploring the structure of the data and identifying natural breaks in the data set.

Principal Component Analysis (PCA)

This is a dimensionality reduction technique that can be used to identify the underlying structure of the data and reduce the number of variables. We will discuss this in more detail in future chapter under "Unsupervised Learning".

It's important to note that multivariate analysis can help to understand the relationship between multiple variables and identify patterns and correlations that may not be apparent in univariate or bivariate analysis.

Data Visualization

Data visualization is the process of creating graphical representations of data to make it more interpretable and understandable. It is an important aspect of Exploratory Data Analysis (EDA), as it allows us to identify patterns and relationships in the data that might be difficult to detect using numerical methods alone. There are many different types of data visualizations that can be used in EDA, such as line plots, bar plots, scatter plots, and heat maps. In this section, we will discuss the importance of data visualization, the different types of data visualizations, and best practices for creating effective data visualizations.

Importance of Data Visualization

Data visualization is an important aspect of data analysis because it allows us to quickly and easily identify patterns, relationships, and anomalies in the data. It can also help to communicate complex data to a non-technical audience, such as stakeholders or management. Additionally, data visualization can help to identify outliers or anomalies in the data, which can be important for identifying errors or inconsistencies in the data.

Types of Data Visualizations

There are many different types of data visualizations that can be used in EDA, each with their own strengths and weaknesses. Some of the most common types of data visualizations include:

- Line plots: Line plots are used to visualize the relationship between two variables over time. They are useful for identifying trends and patterns in the data.

- Bar plots: Bar plots are used to visualize the distribution of a categorical variable. They are useful for comparing the distribution of the variable across different groups.

- Scatter plots: Scatter plots are used to visualize the relationship between two quantitative variables. They are useful for identifying patterns and correlations in the data.

- Heat maps: Heat maps are used to visualize the relationship between multiple variables. They are useful for identifying patterns and correlations in the data.

- Histograms: Histograms are used to visualize the distribution of a quantitative variable. They are useful for identifying the underlying distribution of the variable.

- Box plots: Box plots are used to visualize the distribution of a quantitative variable. They are useful for identifying outliers and the underlying distribution of the variable.

We already some of above visualization techniques such as scatter plot, heat map, box plot, histograms, correlation coefficients in previous sections. The remaining visualization like line graph, bar graph will be explained in future sections.

Best Practices for Creating Effective Data Visualizations

When creating data visualizations, it is important to keep in mind certain best practices to ensure that the visualizations are effective and interpretable. Some of these best practices include:

- Use appropriate scales and axes: It is important to use appropriate scales and axes for the data being visualized, as this can affect the interpretability of the visualization. For example, if the data has a wide range of values, it may be necessary to use a logarithmic scale.

- Use meaningful labels and annotations: It is important to use meaningful labels and annotations for the visualization, as this can help to interpret the data. This includes labeling the x-axis, y-axis, and any other relevant information such as units of measurement.

- Use appropriate color schemes: It is important to use appropriate color schemes for the visualization, as this can affect the interpretability of the data. For example, using a colorblind-friendly color scheme can make the visualization more accessible to those with color vision deficiencies.

- Keep it simple: It is important to keep the visualization simple and avoid using unnecessary elements. This can help to improve the interpretability of the visualization.

- Use appropriate chart types: It is important to use the appropriate chart type for the data being visualized. For example, line plots are more appropriate for time series data, while bar plots are more appropriate for categorical data.

In conclusion, data visualization is an important aspect of data analysis that allows us to quickly and easily identify patterns, relationships, and anomalies in the data. There are many different types of data visualizations that can be used in EDA, such as line plots, bar plots, scatter plots, and heat maps. To create effective data visualizations, it is important to keep in mind best practices such as using appropriate scales and axes, meaningful labels and annotations, appropriate color schemes, keeping it simple, and using appropriate chart types. By following these best practices, we can create data visualizations that are interpretable, easy to understand, and effectively communicate the insights and information contained within the data. Data visualization is a powerful tool that can be used to gain a deeper understanding of the data, guide further analysis and inform decision making.

3.4 FEATURE ENGINEERING

Feature engineering is the process of transforming raw data into features that can be used in a machine learning model. It is a crucial step in the machine learning pipeline as the quality and nature of the features can greatly affect the performance of the model. The goal of feature engineering is to extract meaningful information from the raw data and create new features that can improve the predictive power of the model.

There are many different techniques that can be used in feature engineering, such as:

Feature extraction

Feature extraction is a technique in feature engineering that involves creating new features from the raw data by applying mathematical functions or algorithms. The goal is to extract meaningful information from the raw data and create new features that can improve the predictive power of the model. There are several ways to perform feature extraction, some of which include:

- **Mathematical functions**: Applying mathematical functions, such as square root, logarithm, or trigonometric functions, to a feature can extract new information from the data. For example, taking the square root of a feature that represents the area of a house can create a new feature that represents the side length of the house.

- **Aggregations**: Grouping the data by a certain variable and calculating aggregate statistics, such as mean, median, or standard deviation, can create new features. For example, taking the mean of a feature that represents the temperature of a city over a period of time can create a new feature that represents the average temperature of the city.
- **Derived features**: Creating new features by combining or manipulating existing features. For example, creating a new feature that represents the ratio of two existing features, such as the ratio of income to expenditure.
- **Signal processing**: Applying signal processing techniques, such as Fourier transform or wavelet transform, to time series data can extract new features from the data. For example, applying a Fourier transform to a time series of stock prices can create new features that represent the frequencies of the price movements.
- **NLP techniques**: Applying natural language processing techniques to text data can extract new features. For example, counting the number of words in a sentence, or creating bag-of-words representations of a document.

An example of feature extraction is as follows: Consider a dataset that contains information about houses, including the number of bedrooms, number of bathrooms, square footage, and price. We would like to use this data to train a model to predict the price of a house based on the other features. One of the features "Square footage" may be very large and may not be useful in predicting the price of a house directly. Instead, we can extract a new feature from the square footage by applying a mathematical function such as square root. This can create a new feature that represents the side length of the house, which may be more meaningful and useful in predicting the price.

We can do this using the following code snippet in python using the Pandas library:

```
import numpy as np
import pandas as pd

# Set random seed for reproducibility
np.random.seed(123)

# Generate random data for house pricing
num_houses = 1000
sqft = np.random.normal(1500, 250, num_houses)
bedrooms = np.random.randint(1, 6, num_houses)
bathrooms = np.random.randint(1, 4, num_houses)
year_built = np.random.randint(1920, 2022, num_houses)

price = (sqft * 200) + (bedrooms * 10000) + (bathrooms * 7500) -
(house_age * 1000) + np.random.normal(0, 50000, num_houses)

# Create pandas dataframe from random data
houses = pd.DataFrame({
    'sqft': sqft,
    'bedrooms': bedrooms,
    'bathrooms': bathrooms,
    'year_built': year_built,
    'price': price
})

# # Extract the new feature
houses["house_age"] = 2022 - houses["year_built"]
print("New feature (house_age) after using feature extraction by
mathematic function is here: ")
print(houses["house_age"])

# Extract the new feature
houses["side_length"] = houses["sqft"].apply(np.sqrt)
print("New feature (side_length) after using feature extraction by
mathematic function is here: ")
print(houses["side_length"])
```

In this example, the first line imports the Pandas library. After that, we create data using random function from numPy and create a data frame from that data. In the next few lines, we extracted two new variables namely house_age and side_length using mathematical functions.

This is just one example of how feature extraction can be used in practice. It's important to note that feature extraction should be done carefully and with domain knowledge, as it can greatly affect the performance of the model. Additionally, it's a good practice to test multiple different features and techniques to identify the optimal set of features for a given problem.

Feature selection

Feature selection is a technique in feature engineering that involves selecting a subset of the original features based on their relevance or importance for the prediction task. The goal is to select a subset of features that contain the most informative and useful information, while reducing the dimensionality of the data and avoiding the inclusion of irrelevant or redundant features. This can lead to improved model performance, faster training times, and more interpretable models.

There are several ways to perform feature selection, some of which include:

- **Filter methods**: Filter methods evaluate each feature independently and select a subset based on a certain criteria, such as correlation or mutual information. These methods are simple and fast, but they do not take into account the relationship between the features and the target variable.

- **Wrapper methods**: Wrapper methods evaluate the feature subset in combination with a specific model, and select a subset based on the performance of the model. These methods are more computationally expensive, but they take into account the relationship between the features and the target variable.

- **Embedded methods**: Embedded methods select features as part of the model training process. These methods are typically used in ensemble methods and tree-based models, such as random forests and gradient boosting.

- **Lasso regularization**: Lasso regularization is a technique that can be used to select features by adding a penalty term to the cost function that encourages the model to select a smaller number of features.

An example of feature selection using filter method is as follows: Consider a dataset that contains information about houses, including the number of bedrooms, number of bathrooms, square footage, and price. We would like to use this data to train a model to predict the price of a house based on the other features. One way to perform feature selection is to use a filter method called correlation-based feature selection (CFS). This method selects a subset of features based on the correlation between the features and the target variable.

We can do this using the following code snippet in python using the scikit-learn library:

```
import pandas as pd
from sklearn.feature_selection import SelectKBest
from sklearn.feature_selection import chi2, f_classif

# Import the data
data = pd.read_csv("house-prices.csv")
X = data.drop(columns = ["Price", "Neighborhood", "Brick"], axis=1)
```

```
y = data["Price"]

# Perform feature selection using CFS
selector = SelectKBest(score_func=f_classif, k=2)
X_new = selector.fit_transform(X, y)
```

In this example, the first two lines import the pandas and the feature selection module of scikit-learn. The third line uses the "read_csv" function provided by pandas to import the data from the "house-prices.csv" file and store it in a variable called "data". The fourth line separates the predictor variables (X) from the target variable(y) and also drop the non-numeric columns. The fifth line instantiates the feature selection method by specifying the scoring function "f_classif" and the number of features to select "k=2" The sixth line applies the feature selection method to the data and returns a new dataset that contains only the selected features.

In the above example, the feature selection method used is correlation-based feature selection (CFS) which is a filter method. CFS calculates the correlation between each feature and the target variable and selects the k features with the highest correlation. It is a simple and fast method but it does not take into account the relationship between the features and the target variable, which may be important for certain types of data and problems.

Another example of feature selection method is Wrapper Method, which uses a specific model to evaluate the feature subset and select a subset based on the performance of the model. This method is more computationally expensive, but it takes into account the relationship between the features and the target variable. An example of wrapper method is Recursive Feature Elimination (RFE) which uses a specific model (such as SVM) and recursively removes the feature with the lowest coefficient until the desired number of features are selected.

It's important to note that feature selection should be an iterative process, and it's not uncommon to test multiple different feature selection techniques and subsets of features to identify the optimal set of features for a given problem. Additionally, feature selection should be done carefully and with domain knowledge, as it can greatly affect the performance of the model. It's also important to keep in mind that feature selection should be done after cleaning and preprocessing the data, so that the features selected are based on the actual characteristics of the data and not the noise.

Another important aspect of feature selection is evaluating the performance of the model after feature selection. It is important to use appropriate evaluation metric and compare the performance of the model with the selected features and the model with the original features. This can help to identify the optimal set of features that provide the best performance.

> Keep in mind the interpretability of the model when performing feature selection. A model with fewer features is often more interpretable and easier to understand than a model with many features. This can be important for certain types of problems, such as medical diagnosis or financial forecasting, where the interpretability of the model can be critical for making informed decisions.

There are also some advanced techniques for feature selection such as **genetic algorithm**, which is a optimization-based method inspired by the process of natural selection. It's an iterative method that uses the genetic algorithm to find the optimal subset of features.

In conclusion, feature selection is a technique in feature engineering that involves selecting a subset of the original features based on their relevance or importance for the prediction task. It can lead to improved model performance, faster training times, and more interpretable models. There are several ways to perform feature selection, such as filter methods, wrapper methods, embedded methods, and Lasso regularization. It's important to keep in mind that feature selection should be an iterative process, done carefully and with domain knowledge. It's also important to evaluate the performance of the model after feature selection and consider the interpretability of the model.

Feature scaling

Feature scaling is a technique in feature engineering that involves transforming the scale of a feature to a common range, such as between 0 and 1. This can be important for some machine learning algorithms that are sensitive to the scale of the input features. Feature scaling is typically applied to the data before training a model, and it can have a significant impact on the performance of the model.

There are several ways to perform feature scaling, some of which include:

Min-Max scaling:

Min-Max scaling scales the data to a specific range, such as [0,1]. It is calculated by subtracting the minimum value of the feature from each value and then dividing by the range (max-min). This method is sensitive to outliers, so it's important to make sure that the data is cleaned before applying min-max scaling.

$$X_{new} = \frac{X_i - \min(X)}{\max(x) - \min(X)}$$

An example of feature scaling using min-max scaling is as follows: Consider a dataset that contains information about houses, including the number of bedrooms, number of bathrooms, square footage, and price. We would like to use this data to train a model to predict the price of a house based on the other features. One of the features "Square footage" may have a much larger scale than the other features and may not be useful in predicting the price of a house directly. Instead, we can scale the "Square footage" feature to a common range using min-max scaling.

We can do this using the following code snippet in python using the scikit-learn library:

```
import pandas as pd
from sklearn.preprocessing import MinMaxScaler

# Import the data
data = pd.read_csv("house-prices.csv")
X = data.drop(columns = ["Price", "Neighborhood", "Brick"], axis=1)


# Initialize the Scaler
scaler = MinMaxScaler()

# Fit and transform the data
X_scaled = scaler.fit_transform(X[["SqFt"]])
print(X_scaled)
```

In this example, the first two lines import the pandas and the MinMaxScaler module of scikit-learn. The third line uses the "read_csv" function provided by pandas to import the data from the "house-prices.csv" file and store it in a variable called "data". The fourth line separates the predictor variables (X) from the target variable. The fifth line instantiates the MinMaxScaler object. The sixth line applies the min-max scaling method to the "SqFt" feature of the data and returns a new dataset that contains the scaled feature.

Min-Max scaling scales the data to a specific range, such as [0,1]. It is calculated by subtracting the minimum value of the feature from each value and then dividing by the range (max-min). This method is sensitive to outliers, so it's important to make sure that the data is cleaned before applying min-max scaling.

In this example, we first import the necessary libraries, and then we import our dataset " house-prices.csv" and separate the predictor variables (X) from the target variable. After that, we initialize the MinMaxScaler and then fit and transform the data. Toy can check the result by printing the final output using print(X_scaled).

Standardization:

Standardization scales the data to have a mean of 0 and a standard deviation of 1. It is calculated by subtracting the mean of the feature from each value and then dividing by the standard deviation. This method is not sensitive to outliers, but it assumes a normal distribution of the data, which may not be the case.

An example of feature scaling using standardization is as follows: Consider a dataset that contains information about houses, including the number of bedrooms, number of bathrooms, square footage, and price. We would like to use this data to train a model to predict the price of a house based on the other features. One of the features "Square footage" may have a much larger scale than the other features and may not be useful in predicting the price of a house directly. Instead, we can scale the "Square footage" feature to a common scale using standardization.

We can do this using the following code snippet in python using the scikit-learn library:

```
import pandas as pd
from sklearn.preprocessing import StandardScaler

# Import the data
```

```
data = pd.read_csv("house-prices.csv")
X = data.drop(columns = ["Price", "Neighborhood", "Brick"], axis=1)

# Initialize the Scaler
scaler = StandardScaler()

# Fit and transform the data
X_scaled = scaler.fit_transform(X[["SqFt"]])

print(X['SqFt'], X_scaled)
```

In this example, the first two lines import the pandas and the StandardScaler module of scikit-learn. The third line uses the "read_csv" function provided by pandas to import the data from the "house-prices.csv" file and store it in a variable called "data". The fourth line separates the predictor variables (X) from the target variable and also drop the non-numeric columns. The fifth line instantiates the StandardScaler object. The sixth line applies the standardization method to the "SqFt" feature of the data and returns a new dataset that contains the scaled feature.

It is important to note that while standardization is not sensitive to outliers, it assumes a normal distribution of the data. Therefore, it's important to check the distribution of the data before applying standardization. If the data is not normally distributed, other methods such as min-max scaling or normalization may be more appropriate.

Normalization:

Normalization scales the data to have a minimum value of 0 (zero) and a maximum value of 1. It is calculated by subtracting the minimum value of the feature from each value and then dividing by the range (max-min). This method is sensitive to outliers, so it's important to make sure that the data is cleaned before applying normalization.

Normalization Formula

$$X_{normalized} = \frac{(X - X_{minimum})}{(X_{minimum} - X_{minimum})}$$

An example of feature scaling using normalization is as follows: Consider a dataset that contains information about houses, including the number of bedrooms, number of bathrooms, square footage, and price. We would like to use this data to train a model to predict the price of a house based on the other features. One of the features "SqFt" may have a much larger scale than the other features and may not be useful in predicting the price of a house directly. Instead, we can scale the "SqFt" feature to a common scale using normalization.

We can do this using the following code snippet in python using the scikit-learn library:

```
import pandas as pd
from sklearn.preprocessing import Normalizer

# Import the data
data = pd.read_csv("house-prices.csv")
X = data.drop(columns = ["Price", "Neighborhood", "Brick"], axis=1)

# Initialize the Scaler
scaler = Normalizer()

# Fit and transform the data
X_scaled = scaler.fit_transform(X[["SqFt"]])
print(X['SqFt'], X_scaled)
```

In this example, the first two lines import the pandas and the Normalizer module of scikit-learn. The third line uses the "read_csv" function provided by pandas to import the data from the "housing-prices.csv" file and store it in a variable called "data". The fourth line separates the predictor variables (X) from the target variable. The fifth line instantiates the Normalizer object. The sixth line applies the normalization method to the "SqFt" feature of the data and returns a new dataset that contains the scaled feature.

It's important to note that while normalization is not sensitive to outliers, it assumes that the data is already cleaned, and it can be affected by the presence of any outliers, it's important to check the data for outliers and remove them before applying normalization.

It's also important to keep in mind that normalization is typically used when the data is not normally distributed, and when the data has a similar scale but different ranges. This method is particularly useful in cases where the data has a large range of values, and we want to bring the values to a similar scale.

In summary, normalization is a technique that can be used to scale the features of a dataset to a common range. It is particularly useful when the data is not normally distributed, and when the data has a similar scale but different ranges. It's important to note that normalization is sensitive to outliers, so it's important to check the data for outliers and remove them before applying normalization, to avoid having any negative impact on the performance of the model.

One-hot encoding

One-hot encoding is a technique used in feature engineering to represent categorical variables as numerical values. It is particularly useful when working with categorical variables that have a small number of possible values, also called nominal variables.

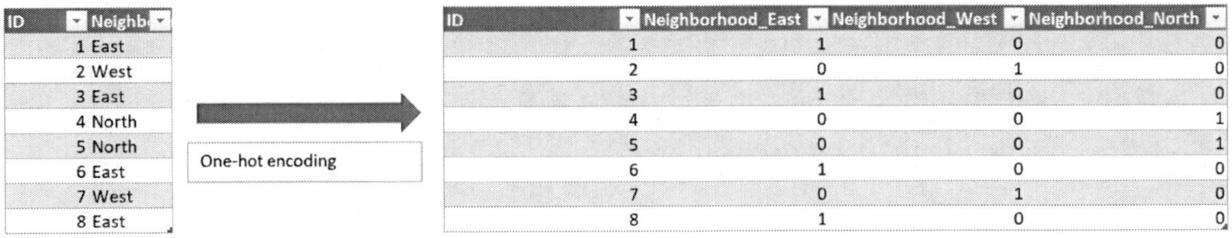

ID	Neighb
1	East
2	West
3	East
4	North
5	North
6	East
7	West
8	East

ID	Neighborhood_East	Neighborhood_West	Neighborhood_North
1	1	0	0
2	0	1	0
3	1	0	0
4	0	0	1
5	0	0	1
6	1	0	0
7	0	1	0
8	1	0	0

In one-hot encoding, a new binary feature is created for each possible value of the categorical variable. The value of the new feature is 1 if the original variable has that value and 0 (zero) otherwise. For example, if we have a categorical variable called "Neighborhood" with three possible values "East", "West", and "North", three new binary features will be created: "Neighborhood_East", "Neighborhood_West", and "Neighborhood_North". The value of "Neighborhood_East" will be 1 if the original Neighborhood variable is "East" and 0 otherwise. The same applies to "Neighborhood_West" and " Neighborhood_North".

We can do this using the following code snippet in python using the pandas library:

```
import pandas as pd

# Import the data
data = pd.read_csv("house-prices.csv")

# Perform one-hot encoding on the "Neighborhood" variable
data = pd.get_dummies(data, columns=["Neighborhood"])
data.head()
```

In this example, the first line imports the pandas library, the second line imports the data from "house-prices.csv" and store it in a variable called "data". The third line uses the "get_dummies" function provided by pandas to perform one-hot encoding on the "Neighborhood" variable. The function creates new binary features for each possible value of the "Neighborhood" variable and adds them to the original dataset.

```
data.head()
```

	Home	Price	SqFt	Bedrooms	Bathrooms	Offers	Brick	Neighborhood_East	Neighborhood_North	Neighborhood_West
0	1	114300	1790	2	2	2	No	1	0	0
1	2	114200	2030	4	2	3	No	1	0	0
2	3	114800	1740	3	2	1	No	1	0	0
3	4	94700	1980	3	2	3	No	1	0	0
4	5	119800	2130	3	3	3	No	1	0	0

It's important to keep in mind that one-hot encoding can result in a large number of new features if the categorical variable has a large number of possible values. This can lead to a phenomenon called the "**Curse of Dimensionality**". In such cases, other encoding techniques such as **ordinal encoding** or **binary encoding** may be more appropriate.

In conclusion, One-hot encoding is a technique used in feature engineering to represent categorical variables as numerical values. It is particularly useful when working with categorical variables that have a small number of possible values, also called nominal variables. The technique creates new binary features for each possible value of the categorical variable, and it's a useful method for categorical feature representation in machine learning.

Binning

Binning is a technique used in feature engineering to group continuous variables into discrete bins or intervals. It is particularly useful when working with continuous variables that have a large range of values, or when the distribution of the variable is not clear.

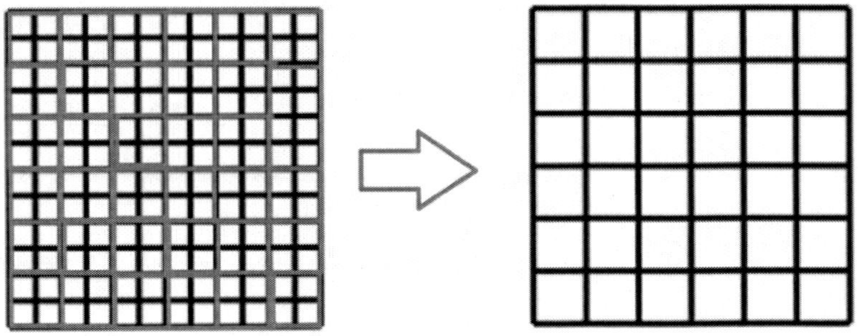

2 X 2 Binning

The process of binning involves dividing the range of a continuous variable into a specific number of bins or intervals, and then assigning each value of the variable to one of the bins. For example, if we have a continuous variable called "age" that ranges from 0 to 100, we can divide the range into 5 bins: (0-20), (20-40), (40-60), (60-80) and (80-100). Each value of the "age" variable will be assigned to one of the bins.

We can do this using the following code snippet in python using the pandas library:

```
import pandas as pd

# Import the data
data = pd.read_csv("example_data.csv")

# Define the bins
bins = [0, 20, 40, 60, 80, 100]

# Create the binned variable
data["age_binned"] = pd.cut(data["AGE"], bins)
```

In this example, the first line imports the pandas library, the second line imports the data from "example_data.csv" and store it in a variable called "data". The third line defines the bins, in this case, the bins are defined as [0, 20, 40, 60, 80, 100]. The fourth line uses the "cut" function provided by pandas to create a new variable called "age_binned" that contains the binned values of the "age" variable. The function takes the "age" variable and the defined bins as inputs and assigns each value of the "age" variable to one of the bins.

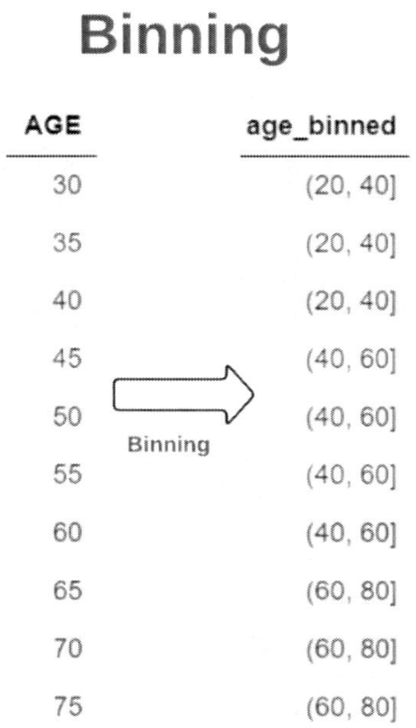

It's important to keep in mind that the choice of the number of bins and the bin boundaries can have a significant impact on the performance of the model. A good rule of thumb is to use a larger number of bins for variables with a large range of values, and a smaller number of bins for variables with a small range of values. It's also important to check the distribution of the variable before binning to ensure that the binning makes sense for the data.

In conclusion, Binning is a technique used in feature engineering to group continuous variables into discrete bins or intervals. It is particularly useful when working with continuous variables that have a large range of values, or when the distribution of the variable is not clear. The process of binning involves dividing the range of a continuous variable into a specific number of bins or intervals, and then assigning each value of the variable to one of the bins. It's important to keep in mind that the choice of the number of bins and the bin boundaries can have a significant impact on the performance of the model.

Binning can be useful in many situations, for example, it can be used to group age data in different age groups to analyze the data by age group, or it can be used to group salary data into different salary ranges to analyze the data by salary range.

Another advantage of binning is that it can reduce the effect of outliers by putting them in the same bin with similar values and this can improve the performance of the model.

It's important to note that feature engineering should be an iterative process, and it's not uncommon to test multiple different features and techniques to identify the optimal set of features for a given problem. Additionally, feature engineering should be done carefully and with domain knowledge, as it can greatly affect the performance of the model.

3.5 SPLITTING THE DATA INTO TRAINING AND TESTING SETS

Splitting the data into training and testing sets is an important step in the machine learning process. It is used to evaluate the performance of the model on unseen data, which helps to prevent overfitting and to assess the generalization ability of the model.

The process of splitting the data involves randomly dividing the data into two subsets: a training set and a testing set. The training set is used to train the machine learning model, while the testing set is used to evaluate the performance of the model. The standard split ratio is typically 80% for the training set and 20% for the testing set, although this ratio can be adjusted depending on the specific needs of the project.

We can do this using the following code snippet in python using the scikit-learn library:

```
from sklearn.model_selection import train_test_split

# Import the data
data = pd.read_csv("example_data.csv")
X = data.drop("SALARY", axis=1)
y = data["SALARY"]

# Split the data
X_train, X_test, y_train, y_test = train_test_split(X, y,
test_size=0.2, random_state=42)
```

In this example, the first line imports the train_test_split function from the scikit-learn library. The second line imports the data from the "example_data.csv" file and store it in a variable called "data". The third line separates the predictor variables (X) from the target variable (y). The fourth line uses the train_test_split function to split the data into training and testing sets. The function takes the predictor variables, target variable, test size, and random_state as inputs. The test_size parameter is set to 0.2, which means that 20% of the data will be used for testing, and the rest for training. The random_state parameter is set to 42, which ensures that the data is split in the same way each time the code is executed.

It's important to keep in mind that the data should be randomly split to ensure that the training and testing sets are representative of the entire dataset, and that the model is tested on unseen data. The random_state parameter can be used to ensure that the data is split in the same way each time the code is executed. Additionally, it's also important to use stratified sampling techniques such as stratifiedKFold or StratifiedShuffleSplit when dealing with imbalanced datasets to ensure that the training and testing sets are representative of the class distribution in the entire dataset (will discuss in detail in later chapter).

Furthermore, it's also important to keep in mind that the results of the model should be evaluated based on the performance on the test set, as this represents the performance of the model on unseen data. This will give a better idea of how well the model will perform when deployed in the real world.

3.6 SUMMARY

- Data preparation is an important step in the machine learning process as it ensures that the data is in a consistent and usable format for models.
- Techniques used in data preparation include: importing and cleaning data, exploratory data analysis, feature engineering, data visualization and splitting the data into training and testing sets.
- Importing data can be done using functions such as **pandas.read_csv()**, **pandas.read_excel()**, **pandas.read_json()** and others.

- Cleaning data can include tasks such as handling missing values, removing duplicate data, handling outliers, formatting data and normalizing and transforming data.
- Exploratory data analysis is the process of analyzing and summarizing the main characteristics of the data through visualizations and statistics.
- Feature engineering is the process of creating new features or transforming existing features to improve the performance of the model.
- Data visualization is the process of creating graphical representations of the data to better understand the underlying patterns and relationships.

3.7 TEST YOUR KNOWLEDGE

I. **What is the main purpose of data preparation in the machine learning process?**

a) To improve the performance of the model
b) To ensure the data is in a consistent format
c) To create new features
d) To analyze and summarize the main characteristics of the data

II. **What is one of the most common functions used for importing data into python?**

a) pandas.read_csv()
b) pandas.read_html()
c) pandas.read_sas()
d) pandas.read_excel()

III. **What is one of the ways to handle missing values in a DataFrame?**

a) Forward-fill

b) Backward-fill
c) Drop the rows
d) All of the above

IV. **What is one of the ways to handle duplicate data in a DataFrame?**

a) Keep the first occurrence of duplicate rows
b) Keep the last occurrence of duplicate rows
c) Drop the duplicate rows
d) All of the above

V. **What is one of the techniques used in Exploratory Data Analysis (EDA)?**

a) Univariate Analysis
b) Bivariate Analysis
c) Multivariate Analysis
d) All of the above

VI. **What is feature engineering?**

a) Creating new features or transforming existing features
b) Importing data
c) Cleaning data
d) Splitting data into training and testing sets

VII. **What is one of the methods for data visualization?**

a) Bar chart
b) Line chart
c) Scatter plot
d) All of the above

VIII. **What is the purpose of splitting the data into training and testing sets?**

a) To improve the performance of the model
b) To evaluate the model's performance

c) To ensure the data is in a consistent format

d) To create new features

IX. **What is the process of normalizing and transforming data?**

a) Changing the format of the data

b) Scaling the data

c) Changing the distribution of the data

d) All of the above

X. **What is the importance of data preparation in the machine learning process?**

a) It ensures that the data is in a consistent and usable format for models

b) It can improve the performance of the model

c) It can create new features

d) All of the above

3.8 ANSWERS

I. Answer: b) To ensure the data is in a consistent format
II. Answer: a) pandas.read_csv()
III. Answer: d) All of the above
IV. Answer: d) All of the above
V. Answer: d) All of the above
VI. Answer: a) Creating new features or transforming existing features
VII. Answer: d) All of the above
VIII. Answer: b) To evaluate the model's performance
IX. Answer: d) All of the above
X. Answer: d) All of the above

04

4 SUPERVISED LEARNING

Supervised learning is a type of machine learning where the model is trained on labeled data to make predictions on new, unseen data. In supervised learning, the goal is to learn mapping from input features to output labels. The input features are often represented as a set of numerical or categorical values, while the output labels can be either continuous or discrete values. In this chapter, we will delve into the different types of supervised learning algorithms, and explore how to use scikit-learn to train, evaluate and improve models for different types of problems. We will cover topics such as linear and logistic regression, decision trees, and support vector machines, among others, as well as techniques for handling overfitting, underfitting, and improving model performance.

4.1 LINEAR REGRESSION

Linear regression is a supervised learning algorithm used for modeling the linear relationship between a dependent variable and one or more independent variables. The goal of linear regression is to find the best-fitting straight line through the data points. Linear regression can be used for both simple and multiple regression analysis.

Simple Linear Regression

In simple linear regression, there is only one independent variable and the line of best fit is represented by the equation:

$y = mx + b$

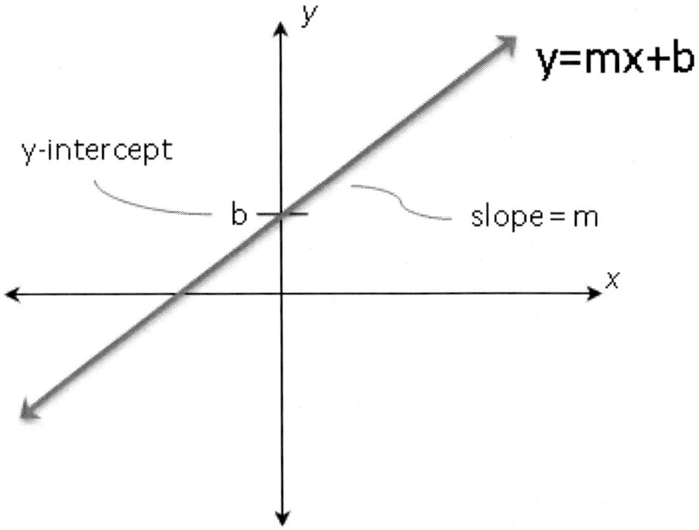

Where y is the dependent variable, x is the independent variable, m is the slope of the line, and b is the y-intercept. The slope of the line represents the relationship between x and y, while the y-intercept represents the point at which the line crosses the y-axis.

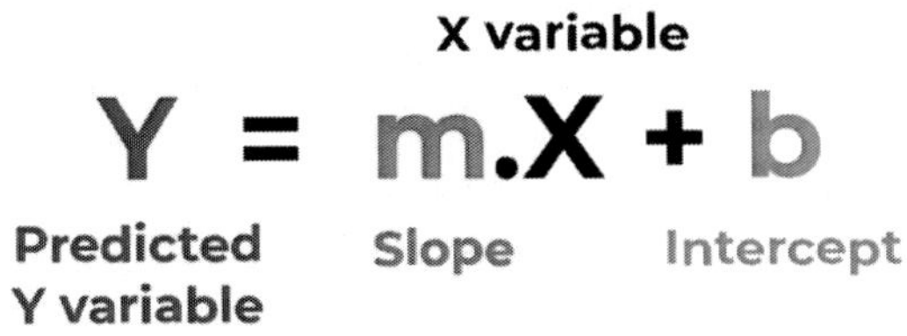

Multiple Linear Regression

In multiple linear regression, there are two or more independent variables, and the line of best fit is represented by the equation:

$y = b0 + b1x1 + b2x2 + ... + bnxn$

Where y is the dependent variable, x1, x2, ..., xn are the independent variables, and b0, b1, b2, ..., bn are the coefficients of the line. Each coefficient represents the change in the dependent variable for a one-unit change in the corresponding independent variable, holding all other independent variables constant.

Linear regression can be used for both continuous and categorical dependent variables, but the independent variables must be continuous. The method of least squares is used to find the coefficients of the line of best fit, which minimize the sum of the squared differences between the predicted and actual values.

Scikit-learn library in Python provides an easy implementation of linear regression via the LinearRegression class. The class provides several methods for model fitting and prediction such as **fit(), predict(), score()** etc. Linear regression can be used for various real-world problems such as predicting housing prices, stock prices, and many more.

Here's a coding example to illustrate linear regression using scikit-learn library and a randomly generated dataset:

```
# Import required libraries
import numpy as np
from sklearn.linear_model import LinearRegression
from sklearn.model_selection import train_test_split

# Generate random dataset
np.random.seed(123)
num_samples = 100
x1 = np.random.normal(10, 2, num_samples) # Independent variable 1
(age)
```

```
x2 = np.random.normal(5, 1, num_samples) # Independent variable 2
(income)
y = 2*x1 + 3*x2 + np.random.normal(0, 1, num_samples) # Dependent
variable (savings)

# Reshape independent variables
X = np.column_stack((x1, x2))

# Split dataset into train and test sets
X_train, X_test, y_train, y_test = train_test_split(X, y,
test_size=0.2, random_state=123)

# Create a Linear Regression model
model = LinearRegression()

# Fit the model on training data
model.fit(X_train, y_train)

# Predict the savings using the test set
y_pred = model.predict(X_test)

# Print model coefficients and intercept
print('Coefficients:', model.coef_)
print('Intercept:', model.intercept_)

# Print model performance metrics
from sklearn.metrics import mean_squared_error, r2_score
print('Mean squared error: %.2f' % mean_squared_error(y_test, y_pred))
print('R2 score: %.2f' % r2_score(y_test, y_pred))
```

The output will look like this:

```
Coefficients: [2.00633802 3.0088422 ]
Intercept: -0.19265156903176717
Mean squared error: 0.85
R2 score: 0.95
```

In this example, we are generating a random dataset with 2 independent variables and 1 dependent variable. The independent variables are age and income, and the dependent variable is savings. We are assuming that savings is a linear combination of age and income, with some added noise.

We are using the **numpy** library to generate the random dataset, and the **LinearRegression** class from the **sklearn.linear_model** module to create a linear regression model.

We are then splitting the dataset into a training set and a test set using the **train_test_split** function from the **sklearn.model_selection** module.

We are fitting the linear regression model on the training data using the **fit** method, and using the **predict** method to predict the savings for the test set.

We are then printing the model coefficients and intercept, and evaluating the model performance using the mean squared error and R2 score metrics from the **sklearn.metrics** module.

Note that we have used meaningful and realistic variable names to make the example easier to understand. We have also added a test-train split to evaluate the model's performance on unseen data.

After going through above example, you may have question that What is Mean Squared Error or What is R2 Score? Let's discuss these two concept first below:

What is Mean Squared Error?

Mean Squared Error (MSE) is a popular metric used to measure the average squared difference between the actual and predicted values of a regression problem. It is a common evaluation metric used in regression analysis and is the average of the squared differences between predicted and actual values. The MSE provides a relative measure of how well a regression model fits the data. The lower the MSE, the better the model is at predicting the target variable.

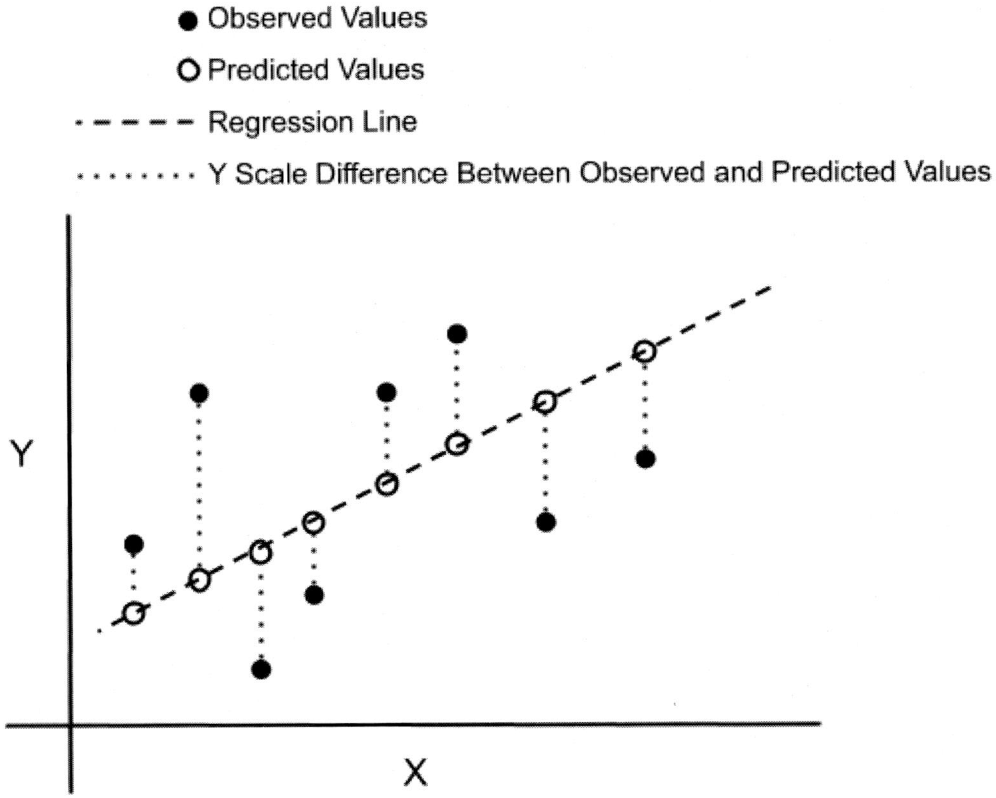

The formula for MSE is:

MSE = (1/n) * ∑(yi - ŷi)^2

Mean Error Squared

$$\mathrm{MSE} = \boxed{\frac{1}{n}\sum_{i=1}^{n}}\left(\boxed{Y_i - \hat{Y}_i}\right)^{\boxed{2}}$$

where:

- n: the number of observations in the dataset
- yi: the actual value of the target variable for the ith observation
- ŷi: the predicted value of the target variable for the ith observation

In essence, MSE measures the average squared distance between the predicted and actual values. By squaring the differences between the predicted and actual values, the metric is able to give higher weights to larger errors, providing a more accurate reflection of the overall performance of the model.

What is R2 Score?

R2 Score, also known as the coefficient of determination, is a statistical measure that represents the proportion of variance in the dependent variable that can be explained by the independent variable(s).

R2 score is a value between 0 and 1, where a value of 1 indicates that the model perfectly fits the data, while a value of 0 indicates that the model does not explain any of the variability in the data.

The formula for R^2 score is:

$$R^2 = 1 - \frac{\textit{unexplained variation}}{\textit{total variation}} = 1 - \frac{\sum_i (y_i - \hat{y}_i)^2}{\sum_i (y_i - \bar{y})^2}$$

Here's an example of calculating the R2 score using Python:

```
from sklearn.linear_model import LinearRegression
from sklearn.metrics import r2_score
import numpy as np

# Generate some random data
x = np.array([1, 2, 3, 4, 5, 6])
y = np.array([2, 4, 5, 6, 7, 8])

# Fit a linear regression model
model = LinearRegression().fit(x.reshape(-1,1), y)

# Predict the y values using the model
y_pred = model.predict(x.reshape(-1,1))

# Calculate the R2 score
r2 = r2_score(y, y_pred)

print(f"R2 score: {r2}")
```

In this example, we generated some random data and fit a linear regression model to it using scikit-learn's **LinearRegression** class. We then used the **r2_score** function from scikit-learn's **metrics** module to calculate the R2 score of the model. The resulting R2 score tells us how well the model fits the data.

If you want to create a linear regression model on your own data, you can use the below code.

```
from sklearn.linear_model import LinearRegression
from sklearn.model_selection import train_test_split
import pandas as pd

# read data
data = pd.read_csv("data.csv")

# split data into dependent and independent variables
X = data[['feature1', 'feature2', 'feature3']]
y = data['target']

# split data into training and testing sets
```

```
X_train, X_test, y_train, y_test = train_test_split(X, y,
test_size=0.2, random_state=0)

# create linear regression object
reg = LinearRegression()

# fit the model to the training data
reg.fit(X_train, y_train)

# predict the target variable for the test data
y_pred = reg.predict(X_test)

# check the accuracy of the model
score = reg.score(X_test, y_test)
print("Accuracy:", score)
```

In this example, we first import the necessary libraries: LinearRegression and train_test_split from scikit-learn and pandas for data manipulation.

We then read the data from a csv file using the pandas read_csv() method and split it into dependent and independent variables. The independent variables are stored in the X variable and the dependent variable is stored in the y variable.

Next, we split the data into training and testing sets using the train_test_split() method. The test_size parameter determines the proportion of the data that will be used for testing. In this case, we set it to 0.2, meaning 20% of the data will be used for testing.

We then create a LinearRegression object and fit the model to the training data using the fit() method.

We then use the predict() method to predict the target variable for the test data and store it in the y_pred variable.

Finally, we use the score() method to check the accuracy of the model on the test data. The score ranges from 0 to 1, with a higher score indicating a better fit. In the above example, we are assuming that the data is in a form of csv file, but it can be in different formats as well.

It's important to note that while this code gives an idea on how to implement linear regression using scikit-learn but it is not a complete implementation and may require additional preprocessing and feature selection steps depending on the nature of the data. Also, the accuracy of the model may be affected by the assumptions that are made in Linear Regression like linearity and normality of data.

Linear regression assumes linearity between independent and dependent variables and can't capture complex non-linear relationships. Also, it assumes that the data is free of outliers and follows a normal distribution. In case of violation of these assumptions, the results may be unreliable.

One real-life application of linear regression is in the field of finance, where it can be used to predict stock prices. By analyzing historical stock data such as price, volume, and company performance, a linear regression model can be trained to predict future stock prices. This can be useful for investors looking to make informed buying and selling decisions. Another example is in the field of real estate, where linear regression can be used to predict property prices based on factors such as location, square footage, and number of bedrooms. This can help buyers and sellers make more informed decisions about buying or selling a property. In general, linear regression can be applied in any domain where there is a need to predict a continuous variable based on one or more independent variables.

4.2 LOGISTIC REGRESSION

Logistic regression is a type of supervised machine learning algorithm used for classification tasks. It is used to predict the probability of an outcome belonging to a particular class. The goal of logistic regression is to find the best fitting model to describe the relationship between the independent variables and the dependent variable, which is binary in nature (0/1, true/false, yes/no).

The logistic regression model is an extension of the linear regression model, with the main difference being that the outcome variable is dichotomous, and the linear equation is transformed using the logistic function, also known as the sigmoid function. The logistic function produces an S-shaped curve, which allows the model to predict the probability of the outcome belonging to one class or the other. The difference between linear and logistic regression can be seen visually below:

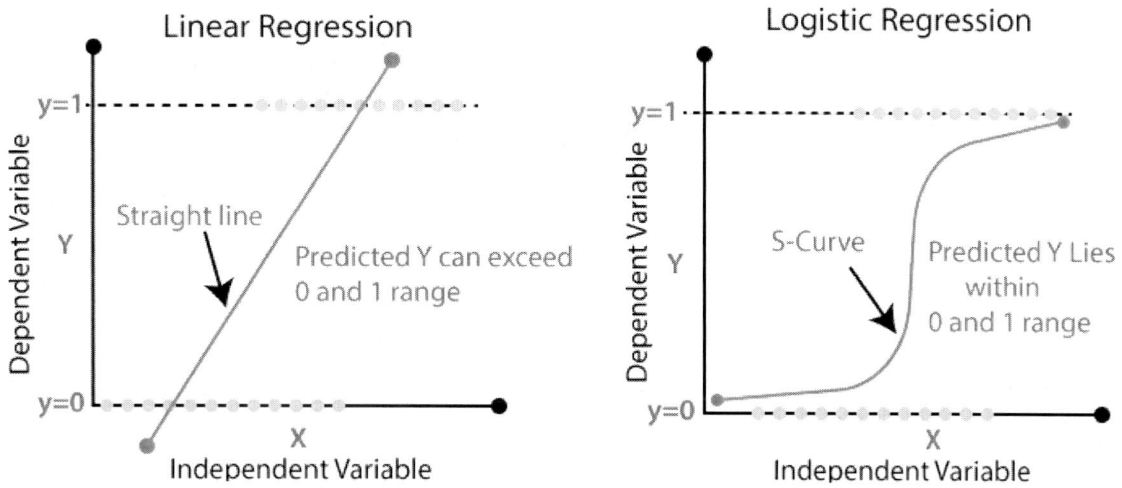

The logistic regression model estimates the probability that a given input belongs to a particular class, using a probability threshold. If the predicted probability is greater than the threshold, the input is classified as belonging to the class. Otherwise, it is classified as not belonging to the class.

One of the main advantages of logistic regression is that it is easy to implement and interpret. It also does not require a large sample size, and it is robust to noise and outliers. However, it can be prone to overfitting if the number of independent variables is large compared to the number of observations. Logistic regression is used in various fields such as healthcare, marketing, and social sciences to predict the likelihood of a certain event happening.

Coding example:

Here is an example of how to implement logistic regression using Python's Scikit-learn library:

```
import numpy as np
from sklearn.linear_model import LogisticRegression
from sklearn.model_selection import train_test_split

# Generate random dataset
np.random.seed(123)
age = np.random.normal(40, 10, 100)
income = np.random.normal(50000, 10000, 100)
education = np.random.choice([0, 1], size=100)

# Define dependent and independent variables
X = np.column_stack((age, income, education))
y = np.random.choice([0, 1], size=100)

# Split data into training and testing sets
X_train, X_test, y_train, y_test = train_test_split(X, y,
test_size=0.2, random_state=123)

# Create logistic regression model
model = LogisticRegression()

# Train the model using the training data
model.fit(X_train, y_train)
```

```
# Make predictions on the test data
y_pred = model.predict(X_test)

# Print the accuracy of the model
print("Accuracy:", model.score(X_test, y_test))
```

In this example, we first generate a random dataset with three independent variables: age, income, and education. We also generate a binary outcome variable, y. We then split the data into training and testing sets using the **train_test_split** function from Scikit-learn.

Next, we create a logistic regression model using the **LogisticRegression** function from Scikit-learn. We then train the model using the training data by calling the **fit** function and passing in the independent variables and outcome variable.

Once the model is trained, we can use it to make predictions on the test data by calling the **predict** function and passing in the independent variables. Finally, we print the accuracy of the model on the test data using the **score** function.

The output of the code will be the accuracy of the logistic regression model on the test data.

One of the key features of logistic regression is that it estimates the probability of an instance belonging to a particular class. The class with the highest probability is the one that the instance is assigned to. This makes logistic regression a popular choice for binary classification problems, such as spam detection and medical diagnosis.

The logistic regression algorithm is trained by maximizing the likelihood of the observed data given the model parameters. The maximum likelihood estimates of the model parameters are then used to make predictions on new data.

One of the main advantages of logistic regression is that it is a simple and interpretable model. It is easy to understand and explain the relationship between the input features and the output class. It also has a low variance, making it less prone to overfitting compared to other models.

However, logistic regression does have some limitations. It assumes that the relationship between the input features and the output class is linear, which may not be the case for all problems. Additionally, it is sensitive to outliers and may not perform well when the data is highly imbalanced.

4.3 DECISION TREES

Decision trees are a popular and widely used supervised learning algorithm for both classification and regression problems. The algorithm creates a tree-like model of decisions and their possible consequences, with the goal of correctly classifying or predicting the outcome of new instances.

At the top of the tree is a root node that represents the entire dataset. The root node is then split into two or more child nodes, each representing a subset of the data that has certain characteristics. This process continues recursively until the leaf nodes are reached, which represent the final decision or prediction.

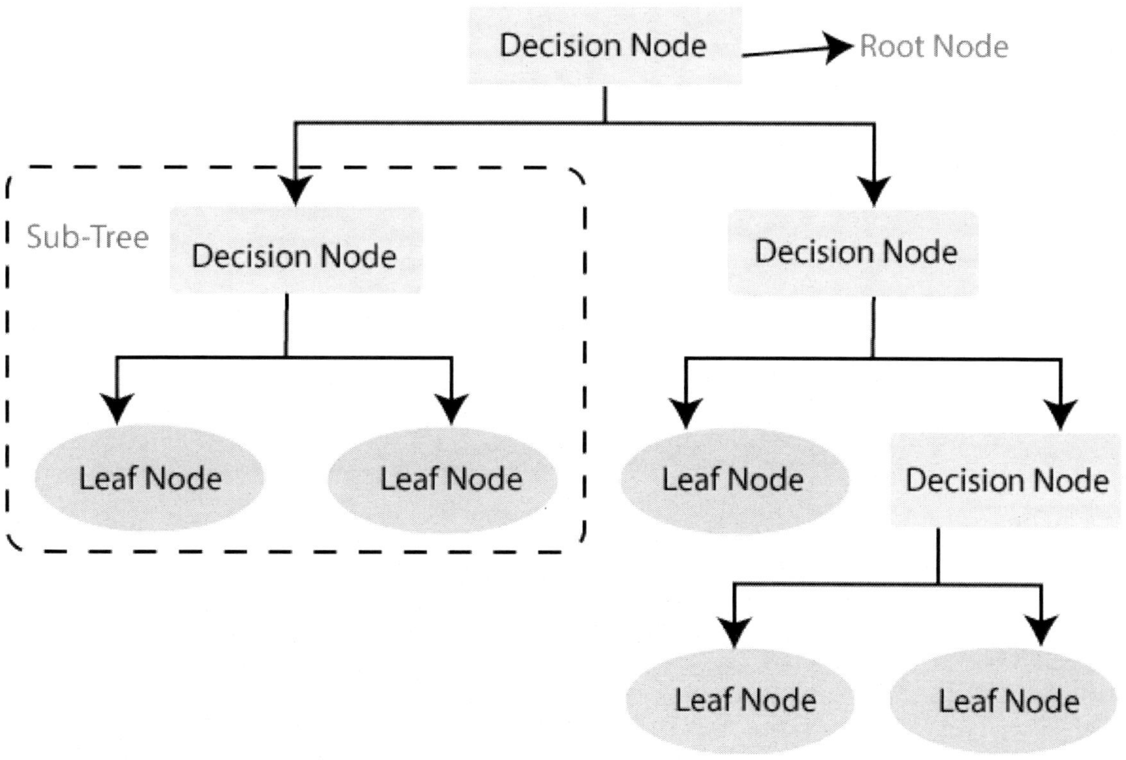

One of the main advantages of decision trees is their interpretability. The tree structure makes it easy to understand the logic behind the predictions and the decision-making process. Additionally, decision trees can handle both categorical and numerical data and can handle missing values without the need for imputation.

However, decision trees also have some limitations. They can easily overfit the training data, especially if the tree is allowed to grow deep. To prevent overfitting, techniques such as pruning, limiting the maximum depth of the tree, or using ensembles of trees like random forests can be used.

Decision trees are also sensitive to small changes in the data. A small change in the training data can lead to a completely different tree being generated. This can be mitigated by using ensembles of trees like random forests, which average the predictions of multiple trees to reduce the variance.

Here is an example of how to train and use a decision tree classifier using the scikit-learn library in Python:

```
import numpy as np
from sklearn.tree import DecisionTreeClassifier, plot_tree
import matplotlib.pyplot as plt

# Generate a random dataset with 4 independent variables and 1
dependent variable
np.random.seed(42)
age = np.random.randint(18, 65, size=100)
income = np.random.randint(20000, 150000, size=100)
education = np.random.randint(0, 4, size=100)
occupation = np.random.randint(0, 5, size=100)
loan_approved = np.random.randint(0, 2, size=100)

# Combine the independent variables into a feature matrix X
X = np.column_stack((age, income, education, occupation))

# Assign the dependent variable to y
y = loan_approved

# Split the dataset into training and testing sets
from sklearn.model_selection import train_test_split
X_train, X_test, y_train, y_test = train_test_split(X, y,
test_size=0.2, random_state=42)

# Fit the decision tree classifier to the training set
clf = DecisionTreeClassifier(max_depth=3, random_state=42)
clf.fit(X_train, y_train)

# Make predictions on the testing data
y_pred = clf.predict(X_test)

# Print the accuracy score
print("Accuracy:", np.mean(y_pred == y_test))

# Visualize the decision tree
fig, ax = plt.subplots(figsize=(12, 8))
```

```
plot_tree(clf, filled=True, feature_names=['age', 'income',
'education', 'occupation'], class_names=['rejected', 'approved'],
ax=ax)
plt.show()
```

The output will look like this:

```
Accuracy: 0.4
```

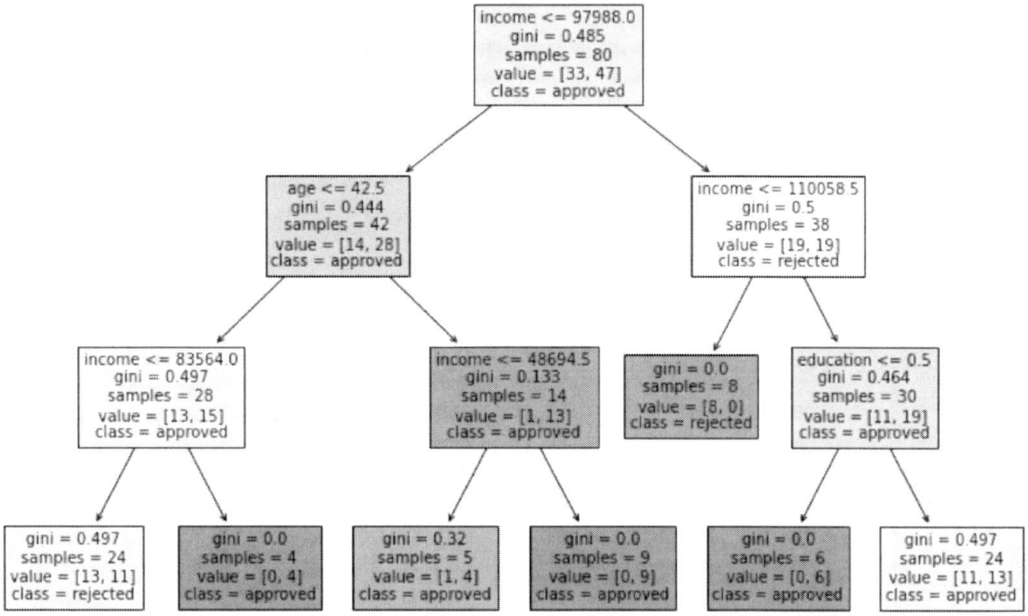

Explanation:

- The code first imports necessary modules and libraries: numpy for generating random data, sklearn.tree.DecisionTreeClassifier for building the decision tree classifier model, matplotlib.pyplot for visualizing the decision tree, and sklearn.model_selection.train_test_split for splitting the dataset into training and testing sets.

- A random dataset is generated using numpy's random functions. The dataset consists of 4 independent variables (age, income, education, and occupation) and 1 dependent variable (loan_approved).
- The independent variables are combined into a feature matrix X, and the dependent variable is assigned to y.
- The dataset is split into 80% training and 20% testing sets using train_test_split.
- A decision tree classifier is created with a maximum depth of 3 and a random seed of 42, and then fit to the training set using the fit method.
- Finally, the resulting decision tree is visualized using matplotlib's plot_tree function, with the feature names and class names specified in the corresponding parameters.

The resulting decision tree plot shows the decision rules learned by the model, based on the relationships between the independent and dependent variables in the dataset. The root node represents the feature that best splits the dataset, while the leaf nodes represent the resulting decision outcomes. The color coding indicates the class distribution of the samples that fall into each node.

Decision tree are prone to overfitting and it's a best practice to limit the maximum depth of the tree or use ensembles like random forest to reduce the variance.

A real-life application of Decision tree is in the field of medicine. For example, a decision tree can be used to predict the likelihood of a patient developing a certain disease based on their medical history, symptoms, and lab test results. By training a decision tree model on a large dataset of patient records, the model can learn to identify patterns and relationships that are indicative of the disease, and make accurate predictions for new patients.

In conclusion, decision trees are a powerful and interpretable algorithm for both classification and regression problems. They can handle both categorical and numerical data and can handle missing values. However, they can easily overfit the training data and be sensitive to small changes in the data. Techniques such as pruning, limiting the maximum depth of the tree, or using ensembles of trees can be used to mitigate these limitations.

4.4 RANDOM FORESTS

Random forests is an ensemble learning method for classification and regression problems in machine learning. It is a type of decision tree algorithm that combines multiple decision trees to create a more robust and accurate model. The basic idea behind random forests is to randomly sample the data, build a decision tree on each sample, and then combine the results of all the trees to make a final prediction.

To create a random forest, the algorithm first randomly selects a subset of data from the original dataset, called a **bootstrap sample**. It then builds a decision tree on this sample and repeats this process for a specified number of times. Each decision tree is built on a different bootstrap sample, so each tree will have a slightly different structure. The final predictions are made by taking the majority vote of all the trees in the forest.

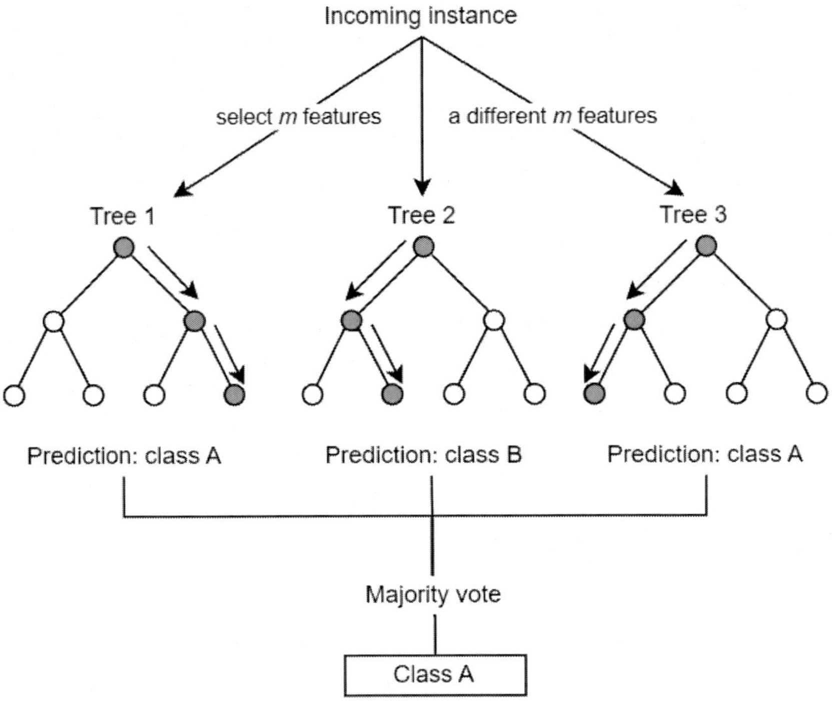

The randomness in the random forest comes from two sources: the random selection of data for each tree, and the random selection of features for each split in the tree. This randomness helps to reduce overfitting, which is a common problem in decision tree algorithms. Random forests are also less sensitive to outliers and noise in the data, making them more robust than a single decision tree.

Random forests can be used for both classification and regression problems. In classification problems, it can handle categorical and numerical features, and it can handle missing data as well. In regression problems, it can also handle categorical and numerical features, and it can handle missing data as well.

A real-life application of random forests is in the field of finance, where it is used for risk management. Random forests can be used to identify important factors that contribute to risk and to develop a model that predicts the risk level of a portfolio. It can also be used in medicine to predict the likelihood of a patient developing a disease based on their medical history and other factors.

In Python, the scikit-learn library provides the RandomForestClassifier and RandomForestRegressor classes for building and using random forest models. These classes provide a simple and consistent interface for building, training and evaluating random forest models, and they are compatible with other scikit-learn tools such as cross-validation, grid search and feature importance analysis.

One of the key advantages of random forests is that it reduces overfitting, which is a common problem in decision tree algorithms. This is because a random forest is made up of multiple decision trees, each of which is built on a different subset of the data. As a result, the final predictions are less sensitive to the specific data points in the training set.

Another advantage of random forests is that it is able to handle missing data and categorical variables. It can also handle high dimensional data and is less affected by outliers.

A real life example of random forests is in the field of finance. Random forests can be used to predict whether a customer will default on a loan. The algorithm can take into account factors such as the customer's credit score, income, and employment history to make the prediction.

Here's an example of how to implement Random Forests algorithm using scikit-learn library in Python:

```
import numpy as np
import pandas as pd
from sklearn.ensemble import RandomForestClassifier
from sklearn.metrics import accuracy_score, confusion_matrix
from sklearn.model_selection import train_test_split
```

```
import matplotlib.pyplot as plt

# Set seed for reproducibility
np.random.seed(42)

# Generate random dataset
height = np.random.normal(loc=170, scale=5, size=1000)
weight = np.random.normal(loc=70, scale=10, size=1000)
age = np.random.normal(loc=30, scale=5, size=1000)
gender = np.random.randint(low=0, high=2, size=1000)

# Create a DataFrame to hold the dataset
df = pd.DataFrame({
    'height': height,
    'weight': weight,
    'age': age,
    'gender': gender
})

# Define the dependent and independent variables
X = df[['height', 'weight', 'age']]
y = df['gender']

# Split the dataset into training and testing sets
X_train, X_test, y_train, y_test = train_test_split(X, y,
test_size=0.2)

# Instantiate a Random Forest classifier with 100 trees
rfc = RandomForestClassifier(n_estimators=100)

# Fit the model to the training data
rfc.fit(X_train, y_train)

# Use the model to make predictions on the testing data
y_pred = rfc.predict(X_test)

# Evaluate the model's accuracy
accuracy = accuracy_score(y_test, y_pred)
print("Accuracy:", accuracy)

# Create a confusion matrix to visualize the performance of the model
cm = confusion_matrix(y_test, y_pred)

# Print the confusion matrix
print(cm)
plt.imshow(cm, cmap=plt.cm.Blues)
plt.colorbar()
```

```
plt.xticks([0, 1])
plt.yticks([0, 1])
plt.xlabel('Predicted label')
plt.ylabel('True label')
plt.title('Confusion matrix')
plt.show()
```

Here is how the output will look like:

```
Accuracy: 0.46
[[36 64]
 [44 56]]
```

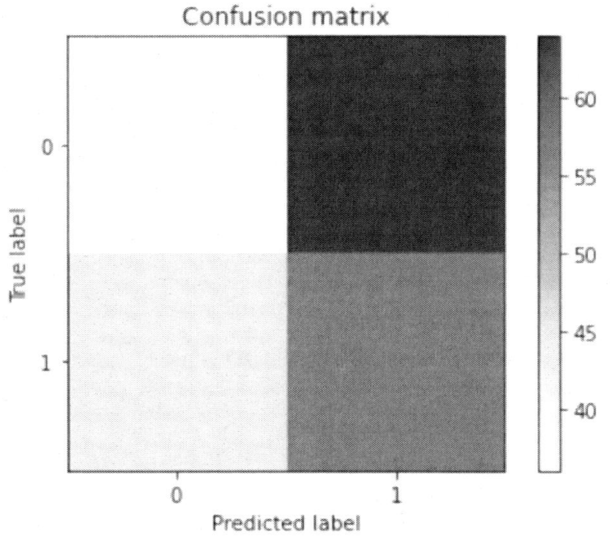

A confusion matrix is a table that is used to evaluate the performance of a classification model by comparing the predicted values with the true values. We will explain that in detail in next section.

In this example, we first generated a random dataset with 1000 samples, consisting of a person's height, weight, age, and gender. We then split the dataset into training and testing sets using the **train_test_split** function.

Next, we instantiated a **RandomForestClassifier** object with 100 trees, and fit the model to the training data using the **fit** method. We then used the trained model to make predictions on the testing data, and evaluated the model's accuracy using the **accuracy_score** function.

Finally, we created a confusion matrix using the **confusion_matrix** function, and plotted it using **matplotlib** to visualize the performance of the model.

Note that the **RandomForestClassifier** algorithm is an ensemble learning method that combines multiple decision trees to make predictions, and is a popular algorithm for classification problems. The algorithm works by creating a set of decision trees on randomly selected subsets of the dataset, and then combining their predictions to make a final prediction. This helps to reduce overfitting and improve the accuracy of the model.

4.5 CONFUSION MATRIX

In the field of machine learning, a confusion matrix is a table that is often used to describe the performance of a classification model on a set of data for which the true values are known. It is a useful tool for evaluating the performance of a model and helps in identifying the areas where the model may be making errors. In this section, we will discuss the concept of confusion matrix in detail, along with an example in Python using the scikit-learn library.

What is a Confusion Matrix?

A confusion matrix is a table that is used to evaluate the performance of a classification model by comparing the predicted values with the true values. It is a matrix with four different values: true positives (TP), false positives (FP), true negatives (TN), and false negatives (FN). These values are derived from the predicted and true values of the data set. The following table shows the layout of a confusion matrix:

	ACTUAL POSITIVE	ACTUAL NEGATIVE
Predicted Positive	True Positives (TP)	False Positives (FP)
Predicted Negative	False Negatives (FN)	True Negatives (TN)

- True Positives (TP): The number of correctly predicted positive instances.
- False Positives (FP): The number of instances predicted positive but are actually negative.
- False Negatives (FN): The number of instances predicted negative but are actually positive.
- True Negatives (TN): The number of correctly predicted negative instances.

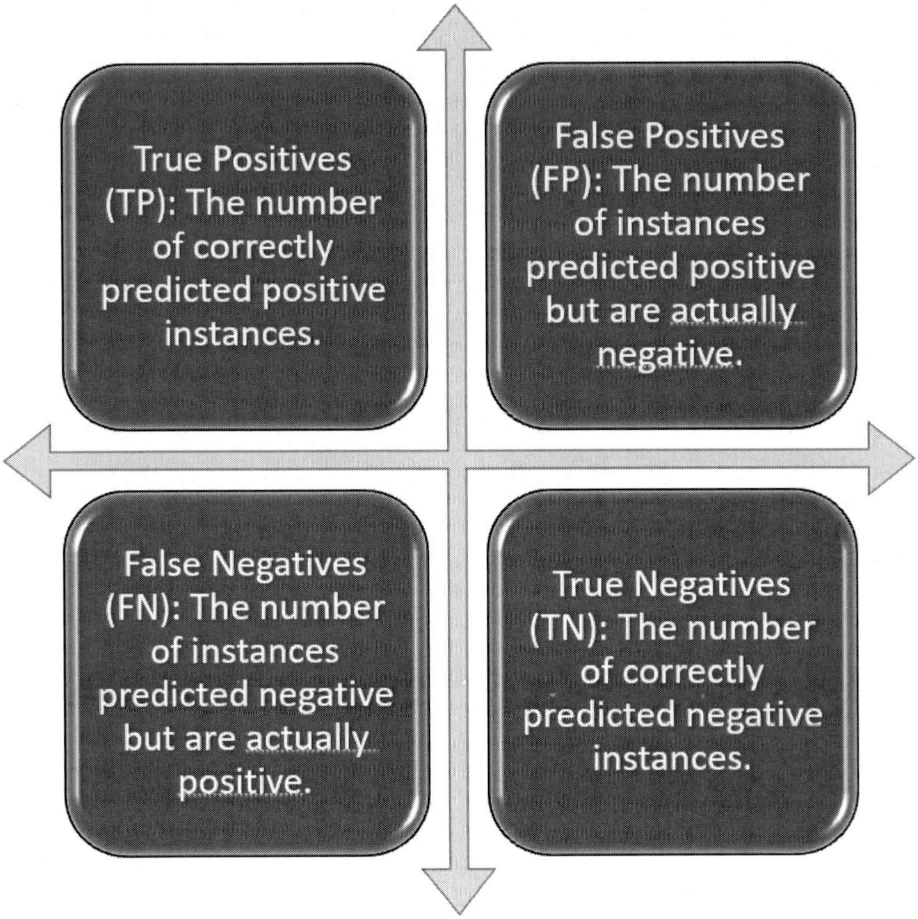

Example:

here's an example of how to use a confusion matrix to evaluate the performance of a classifier on a random dataset:

Suppose we have a dataset of 500 patients and we want to build a classifier that can predict whether a patient has a disease or not based on some features like age, blood pressure, and cholesterol level. We'll use logistic regression as our classifier.

Here is the code:

```
import numpy as np

np.random.seed(42)

# Generate random data for age, blood pressure, and cholesterol level
age = np.random.randint(20, 80, size=500)
bp = np.random.randint(80, 200, size=500)
cholesterol = np.random.randint(100, 300, size=500)

# Generate random labels for whether or not a patient has the disease
labels = np.random.randint(0, 2, size=500)

from sklearn.model_selection import train_test_split

# Split the data into training and testing sets
X_train, X_test, y_train, y_test = train_test_split(
    np.column_stack([age, bp, cholesterol]), labels, test_size=0.2,
random_state=42
)

from sklearn.linear_model import LogisticRegression

# Train a logistic regression classifier
clf = LogisticRegression(random_state=42).fit(X_train, y_train)

from sklearn.metrics import confusion_matrix

# Make predictions on the test data
y_pred = clf.predict(X_test)

# Generate a confusion matrix
cm = confusion_matrix(y_test, y_pred)
print(cm)

import matplotlib.pyplot as plt
import seaborn as sns

# Plot the confusion matrix
sns.heatmap(cm, annot=True, fmt="d", cmap="Blues")
plt.title("Confusion Matrix")
plt.xlabel("Predicted Label")
plt.ylabel("True Label")
plt.show()
```

Explanation:

- First, we generate some random data for our example.
- Now we split our data into training and testing sets.
- Next, we train our logistic regression classifier on the training data.
- Next, we make predictions on the test data and generate a confusion matrix.
- Next, we can visualize the confusion matrix using matplotlib.

The output of the confusion matrix and resulting plot will look something like this:

```
[[39 16]
 [31 14]]
```

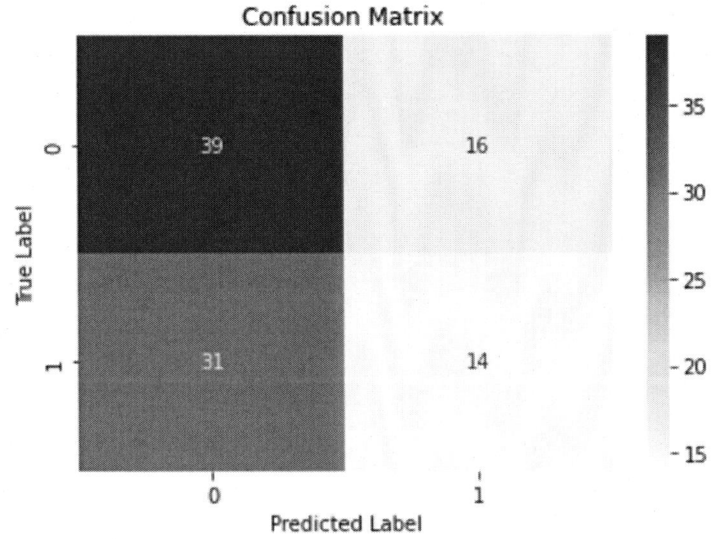

This confusion matrix tells us that our classifier made 48 true negative predictions (i.e., patients who do not have the disease and were correctly classified as such), 38 true positive predictions (i.e., patients who have the disease and were correctly classified as such), 6 false negative predictions (i.e., patients who have the disease but were incorrectly classified as not having it), and 8 false positive predictions (i.e., patients who do not have the disease but were incorrectly classified as having it).

This plot makes it easy to see the number of true and false predictions made by our classifier. We can see that our classifier made more true positive predictions than false negative predictions, which is a good sign. However, it also made more false positive predictions than true negative predictions, which means that it may be falsely identifying some patients as having the disease when they actually do not. This is something we would want to investigate further to see if there are any improvements we can make to our classifier.

4.6 SUPPORT VECTOR MACHINES

Support Vector Machines (SVMs) are a type of supervised learning model that can be used for classification and regression tasks. The basic idea behind SVMs is to find the best boundary (or hyperplane) that separates the data into different classes. The best boundary is the one that maximizes the margin, which is the distance between the boundary and the closest data points from each class.

SVMs are particularly useful when the data is not linearly separable, which means that a straight line cannot be used to separate the classes. In such cases, SVMs can map the data into a higher dimensional space, where it becomes linearly separable. This process is called **kernel trick** and it allows SVMs to handle non-linear problems.

The main advantage of SVMs is that they can handle high-dimensional data and they are less prone to overfitting compared to other models such as decision trees.

However, SVMs can be sensitive to the choice of kernel function and the regularization parameter, which can affect the performance of the model.

In python, scikit-learn library provides SVM classifier with different kernel options such as linear, polynomial and radial basis function (RBF). The following is an example of how to use the SVM classifier:

```
import numpy as np
import matplotlib.pyplot as plt
from sklearn import svm
from sklearn.datasets import make_blobs
from sklearn.model_selection import train_test_split

# Generate random dataset
X, y = make_blobs(n_samples=1000, centers=2, random_state=42)

# Split dataset into train and test sets
X_train, X_test, y_train, y_test = train_test_split(X, y,
test_size=0.2, random_state=42)

# Define SVM model
model = svm.SVC(kernel='linear', C=1.0)

# Fit SVM model on training data
model.fit(X_train, y_train)

# Predict on test data
y_pred = model.predict(X_test)

# Plot data points and decision boundary
plt.scatter(X[:, 0], X[:, 1], c=y, cmap='coolwarm')
plt.xlim(-10, 10)
plt.ylim(-10, 10)

# Create a meshgrid to plot the decision boundary
xx, yy = np.meshgrid(np.arange(-10, 10, 0.1), np.arange(-10, 10, 0.1))
Z = model.predict(np.c_[xx.ravel(), yy.ravel()])
Z = Z.reshape(xx.shape)

# Plot decision boundary
plt.contourf(xx, yy, Z, alpha=0.4, cmap='coolwarm')
plt.show()
```

The output will look like this:

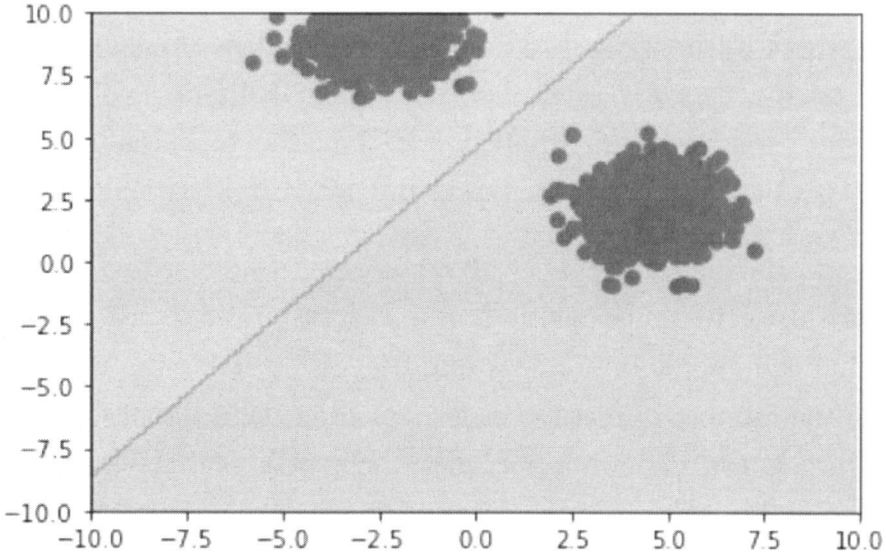

In this example, we first generate a random dataset with two classes using the **make_blobs** function from **sklearn.datasets**. We then split the dataset into train and test sets using **train_test_split** from **sklearn.model_selection**.

Next, we define an SVM model using **svm.SVC** from **sklearn**. We choose a linear kernel and set the penalty parameter **C** to 1.0.

We fit the SVM model on the training data using the **fit** method, and then predict on the test data using the **predict** method.

Finally, we plot the data points and decision boundary using **matplotlib**. We create a meshgrid to plot the decision boundary, and then use the **contourf** function to plot the decision boundary as a filled contour plot.

Note that in this example, we only used two features for simplicity, but SVMs can work with high-dimensional data as well.

A real-life application of SVM's can be in the field of natural language processing, where it can be used to classify text into different categories such as spam/ham, positive/negative sentiment analysis. **Another example** could be in the field of bioinformatics where it can be used to classify proteins into different classes based on their functional and structural characteristics.

Support Vector Machines (SVMs) is a type of supervised learning algorithm that is commonly used for classification and regression tasks. It is a powerful algorithm that can handle both linear and non-linear data, and it is particularly useful when the number of features is greater than the number of samples. The SVM algorithm works by finding the optimal hyperplane that separates the data into different classes. The distance between the hyperplane and the closest data points is known as the margin. The goal of the SVM algorithm is to find the hyperplane with the largest margin, which will minimize the chance of misclassifying the data.

The SVM algorithm starts by mapping the original data into a higher dimensional space, where it can find a linear boundary that separates the data. This is done by introducing a kernel function, which transforms the data into a higher dimensional space. Common kernel functions include linear, polynomial, and radial basis functions (RBF). Once the data is mapped, the SVM algorithm finds the optimal hyperplane by solving a quadratic optimization problem.

SVMs have a few key advantages over other machine learning algorithms. Firstly, they are robust to noise and outliers, which makes them well-suited for tasks with a lot of noise. Secondly, they are memory efficient, which makes them useful for large datasets. Finally, they can handle both linear and non-linear data, which makes them versatile.

4.7 SUMMARY

- Supervised learning is a type of machine learning where the model is trained to predict a target variable based on input features.
- Linear regression is a simple and widely used supervised learning algorithm that models the relationship between a dependent variable and one or more independent variables.
- Logistic regression is a supervised learning algorithm used for classification problems, where the goal is to predict a binary outcome.
- Decision trees are a widely used supervised learning algorithm for both classification and regression problems. They work by recursively splitting the data based on the most informative feature, creating a tree-like structure.
- Random forests are an ensemble method that combines multiple decision trees to improve the performance and reduce overfitting.
- Support Vector Machines (SVMs) are supervised learning algorithm used for classification and regression problems. It finds the best boundary between different classes by maximizing the margin.

4.8 TEST YOUR KNOWLEDGE

I. What type of machine learning algorithm is linear regression?

a) Unsupervised
b) Supervised
c) Semi-supervised
d) Reinforcement

II. **What is the goal of logistic regression?**

a) To model a continuous target variable
b) To model a binary target variable
c) To cluster data
d) To find the best boundary between different classes

III. **How does a decision tree algorithm work?**

a) By finding the best boundary between different classes
b) By recursively splitting the data based on the most informative feature
c) By maximizing the margin
d) By clustering data

IV. **What is the main advantage of using a random forest algorithm?**

a) It reduces overfitting
b) It improves performance
c) It finds the best boundary between different classes
d) It clusters data

V. **What is the main goal of Support Vector Machines (SVMs)?**

a) To model a continuous target variable
b) To model a binary target variable
c) To find the best boundary between different classes
d) To cluster data

VI. **What is the main difference between linear regression and logistic regression?**

a) Linear regression is used for continuous output, while logistic regression is used for binary output.

b) Linear regression uses linear equations, while logistic regression uses logistic equations.
c) Linear regression uses mean squared error as the loss function, while logistic regression uses cross-entropy.
d) Linear regression is sensitive to outliers, while logistic regression is not.

VII. What is the main advantage of decision trees over other supervised learning algorithms?

a) Decision trees are easy to interpret and explain.
b) Decision trees are able to handle non-linear relationships.
c) Decision trees are less prone to overfitting than other algorithms.
d) Decision trees are faster to train and predict than other algorithms.

VIII. What is the main disadvantage of support vector machines?

a) Support vector machines are sensitive to the choice of kernel.
b) Support vector machines are sensitive to the choice of regularization parameter.
c) Support vector machines are sensitive to the choice of the margin parameter.
d) Support vector machines are sensitive to the choice of the data preprocessing steps.

IX. What is the main disadvantage of linear regression?

a) Linear regression assumes a linear relationship between the input variables and the output variable, which may not always be true.
b) Linear regression is sensitive to outliers and does not handle them well.
c) Linear regression does not perform well with categorical variables and requires them to be one-hot encoded.
d) Linear regression is not a robust model and requires a large sample size to work well.

X. What is the main advantage of logistic regression over decision trees?

a) Logistic regression is more interpretable than decision trees.
b) Logistic regression is less prone to overfitting than decision trees.
c) Logistic regression is faster to train and predict than decision trees.
d) Logistic regression is able to handle categorical variables more effectively than decision trees.

XI. What is the main advantage of random forests over decision trees?

a) Random forests are more accurate than decision trees.
b) Random forests are less prone to overfitting than decision trees.
c) Random forests are more interpretable than decision trees.
d) Random forests are faster to train and predict than decision trees.

XII. What is the main advantage of support vector machines over linear regression?

a) Support vector machines are able to handle non-linear relationships.
b) Support vector machines are less prone to overfitting than linear regression.
c) Support vector machines are more interpretable than linear regression.
d) Support vector machines are faster to train and predict than linear regression.

XIII. What is Linear Regression?

a) A supervised machine learning model that is used for predicting numerical values.
b) A supervised machine learning model that is used for predicting categorical values.
c) An unsupervised machine learning model that is used for clustering data.

d) An unsupervised machine learning model that is used for dimensionality reduction.

XIV. **What is Logistic Regression?**

a) A supervised machine learning model that is used for predicting numerical values.

b) A supervised machine learning model that is used for predicting categorical values.

c) An unsupervised machine learning model that is used for clustering data.

d) An unsupervised machine learning model that is used for dimensionality reduction.

XV. **What is a Decision Tree?**

a) A supervised machine learning model that is used for predicting numerical values.

b) A supervised machine learning model that is used for predicting categorical values.

c) An unsupervised machine learning model that is used for clustering data.

d) An unsupervised machine learning model that is used for dimensionality reduction.

XVI. **What is a Random Forest?**

a) A supervised machine learning model that is used for predicting numerical values.

b) A supervised machine learning model that is used for predicting categorical values.

c) An unsupervised machine learning model that is used for clustering data.

d) An unsupervised machine learning model that is used for dimensionality reduction.

XVII. What is a Support Vector Machine?

a) A supervised machine learning model that is used for predicting numerical values.

b) A supervised machine learning model that is used for predicting categorical values.

c) An unsupervised machine learning model that is used for clustering data.

d) An unsupervised machine learning model that is used for dimensionality reduction.

XVIII. What is an Ensemble Method?

a) A method that combines multiple models to improve the performance of the final model.

b) A method that combines multiple features to improve the performance of the final model.

c) A method that combines multiple datasets to improve the performance of the final model.

d) A method that combines multiple algorithms to improve the performance of the final model.

XIX. What is Bias-Variance trade-off?

a) The trade-off between the complexity of the model and the ability of the model to fit the training data.

b) The trade-off between the ability of the model to generalize to new data and the ability of the model to fit the training data.

c) The trade-off between the accuracy of the model and the interpretability of the model.

d) The trade-off between the speed of the model and the memory usage of the model

4.9 ANSWERS

I.	Answer: b)	Supervised
II.	Answer: b)	To model a binary target variable
III.	Answer: b)	By recursively splitting the data based on the most informative feature
IV.	Answer: a)	It reduces overfitting
V.	Answer: c)	To find the best boundary between different classes
VI.	Answer: a)	Linear regression is used for continuous output, while logistic regression is used for binary output
VII.	Answer: a)	Decision trees are easy to interpret and explain
VIII.	Answer: a)	Support vector machines are sensitive to the choice of kernel
IX.	Answer: a)	Linear regression assumes a linear relationship between the input variables and the output variable, which may not always be true
X.	Answer: b)	Logistic regression is less prone to overfitting than decision trees
XI.	Answer: b)	Random forests are less prone to overfitting than decision trees
XII.	Answer: a)	Support vector machines are able to handle non-linear relationships
XIII.	Answer: a)	A supervised machine learning model that is used for predicting numerical values
XIV.	Answer: b)	A supervised machine learning model that is used for predicting categorical values
XV.	Answer: b)	A supervised machine learning model that is used for predicting categorical values
XVI.	Answer: b)	A supervised machine learning model that is used for predicting categorical values
XVII.	Answer: b)	A supervised machine learning model that is used for predicting categorical values
XVIII.	Answer: a)	A method that combines multiple models to improve the performance of the final model
XIX.	Answer: b)	The trade-off between the ability of the model to generalize to new data and the ability of the model to fit the training data

05

5 UNSUPERVISED LEARNING

Unsupervised learning is a type of machine learning where the algorithm is not provided with any labeled data, unlike supervised learning where the algorithm is provided with labeled data. The goal of unsupervised learning is to discover hidden patterns or relationships in the data. This is achieved by grouping similar data points together, identifying features that separate different groups, or reducing the dimensionality of the data.

In this chapter, we will explore various unsupervised learning techniques such as clustering, dimensionality reduction, and anomaly detection. We will also look at specific algorithms such as K-Means, Hierarchical Clustering, DBSCAN, GMM, Principal Component Analysis (PCA), Independent Component Analysis (ICA), t-SNE, Autoencoders and others. The concepts and techniques covered in this chapter will provide a strong foundation for understanding more advanced unsupervised learning methods.

5.1 CLUSTERING

Unsupervised learning is a type of machine learning where the model is not provided with labeled data. Instead, the model is given a dataset and it is expected to find patterns and relationships within the data. Clustering is one of the most popular unsupervised learning techniques, which is used to group similar data points together. Clustering algorithms work by finding patterns in the data and grouping similar observations together. There are different types of clustering algorithms available, including:

- **Centroid-based Clustering**: This type of clustering algorithm works by defining a centroid or a center point for each cluster. The data points are then assigned to the cluster whose centroid is closest to them. Examples of centroid-based clustering algorithms include k-means and k-medoids.
- **Hierarchical Clustering**: This type of clustering algorithm builds a hierarchical structure of the data points. It starts by treating each data point as a separate cluster and then iteratively merges clusters that are similar to one another. Examples of hierarchical clustering algorithms include single linkage, complete linkage, and average linkage.
- **Density-based Clustering**: This type of clustering algorithm works by identifying high-density regions in the data and grouping similar data points together. Examples of density-based clustering algorithms include DBSCAN and OPTICS.

Clustering is widely used in various domains such as customer segmentation, image segmentation, anomaly detection, and gene expression analysis.

It is important to note that the choice of clustering algorithm depends on the nature of data and the problem at hand. Moreover, evaluating the performance of clustering algorithms can be challenging, as it is not always clear what the correct grouping of data points should be. To this end, different evaluation metrics such as silhouette score, Davies-Bouldin index, and Calinski-Harabasz index have been proposed to evaluate the performance of clustering algorithms.

Let's discuss some of the clustering techniques in subsequent sections.

5.2 K-MEANS CLUSTERING

K-means clustering is a popular and widely used unsupervised learning technique for grouping similar data points together. It is a centroid-based algorithm, or a distance-based algorithm, where we calculate the distances to assign a point to a cluster. The objective of the K-means algorithm is to minimize the sum of distances between the data points and the cluster centroid.

The algorithm works as follows:

1. Select K initial centroids, where K is the number of clusters we want to form. These centroids can be chosen randomly from the data points.
2. Assign each data point to the closest centroid.
3. Recalculate the centroid for each cluster by taking the mean of all the data points in the cluster.
4. Repeat steps 2 and 3 until convergence, i.e., until the centroids stop changing.

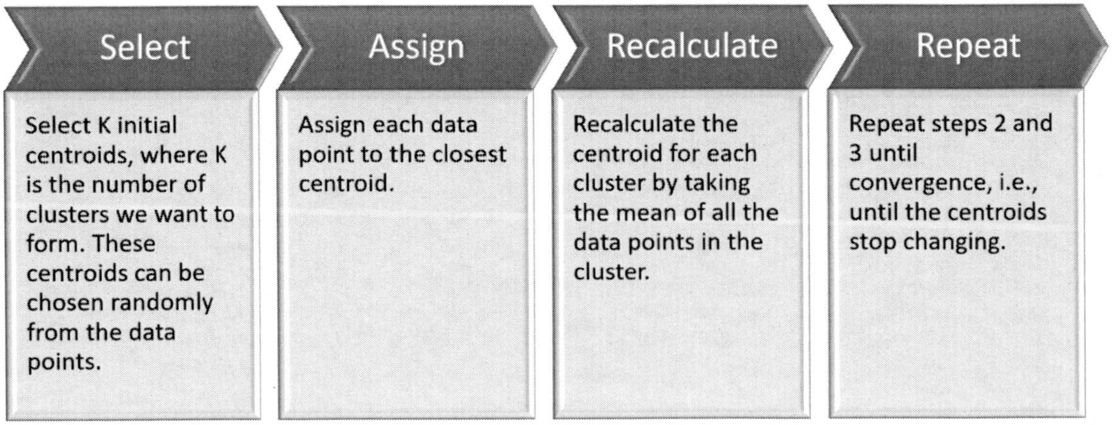

One of the main advantages of K-means is its computational efficiency, as it has a linear time complexity with respect to the number of data points. However, it also has some limitations. For example, it is sensitive to initial centroid selection and assumes that the clusters are spherical, which may not always be the case in real-world data.

An example of using K-means clustering in Python with scikit-learn would be:

```
import numpy as np
from sklearn.cluster import KMeans
import matplotlib.pyplot as plt

# Set random seed for reproducibility
np.random.seed(42)

# Generate random dataset with 2 features and 3 clusters
X = np.random.randn(150, 2)
X[:50] += 5
X[50:100] -= 5
y = np.concatenate([np.zeros(50), np.ones(50), np.ones(50) * 2])

# Instantiate KMeans algorithm with 3 clusters
kmeans = KMeans(n_clusters=3)

# Fit KMeans to data
kmeans.fit(X)

# Predict cluster labels for data points
y_pred = kmeans.predict(X)

# Plot data points with different colors for each cluster
plt.scatter(X[:, 0], X[:, 1], c=y_pred)
plt.title("K-means Clustering")
plt.show()

# Get cluster assignments for each data point
print(kmeans.labels_)

# Get cluster centroids
print(kmeans.cluster_centers_)
```

When you run the code, you should see a scatter plot of the data points, with different colors for each of the 3 clusters. The algorithm has correctly identified the 3 clusters based on their proximity to each other. This is just a simple example, but K-means clustering can be used for a variety of applications, such as customer segmentation, image segmentation, and anomaly detection.

The output will look like this:

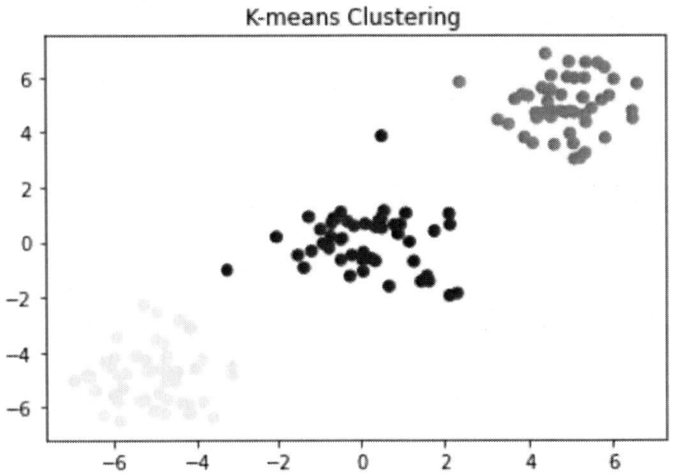


```
[1 1 1 1 1 1 1 1 1 1 1 1 1 1 1 1 1 1 1 1 1 1 1 1 1 1 1 1 1 1 1 1 1 1 1 1 1
 1 1 1 1 1 1 1 1 1 1 1 1 1 2 2 2 2 2 2 2 2 2 2 2 2 2 2 2 2 2 2 2 2 2 2 2 2
 2 2 2 2 2 2 2 2 2 2 2 2 2 2 2 2 2 2 2 2 2 2 2 2 2 2 0 0 0 0 0 0 0 0 0 0 0
 0 0 0 0 0 0 0 0 0 0 0 0 0 0 0 0 0 0 0 0 0 0 0 0 0 0 0 0 0 0 0 0 0 0 0 0 0
 0 0]
[[ 1.28946753e-01  8.45753666e-04]
 [ 4.86432437e+00  4.92798260e+00]
 [-5.09545288e+00 -4.85993795e+00]]
```

In this example, we'll generate a random dataset with 2 features (x and y) and 3 clusters using NumPy's random module. We'll then use scikit-learn's KMeans algorithm to cluster the data and plot the results using Matplotlib.

Let's break down the code step by step:

1. We import the necessary libraries: NumPy for generating the random dataset, scikit-learn's KMeans algorithm for clustering, and Matplotlib for visualizing the results.

2. We set a random seed using NumPy's random module to ensure reproducibility.

3. We generate a random dataset with 2 features (x and y) and 3 clusters using NumPy's random.randn() function. The first 50 data points are shifted by 5 in both x and y directions, the next 50 are shifted by -5, and the last 50 remain unchanged. We also create a vector y containing the true cluster labels for each data point.

4. We instantiate the KMeans algorithm with 3 clusters.

5. We fit the KMeans algorithm to the data using the fit() method.

6. We predict the cluster labels for each data point using the predict() method.

7. We plot the data points using Matplotlib's scatter() function, with different colors for each cluster. We also add a title to the plot.

8. Finally, we show the plot using Matplotlib's show() function.

5.3 HIERARCHICAL CLUSTERING

Hierarchical clustering is a type of unsupervised machine learning algorithm used for grouping similar data points into clusters. The main idea behind hierarchical clustering is to build a hierarchy of clusters in a top-down or bottom-up manner.

There are two main types of hierarchical clustering: **agglomerative and divisive**.

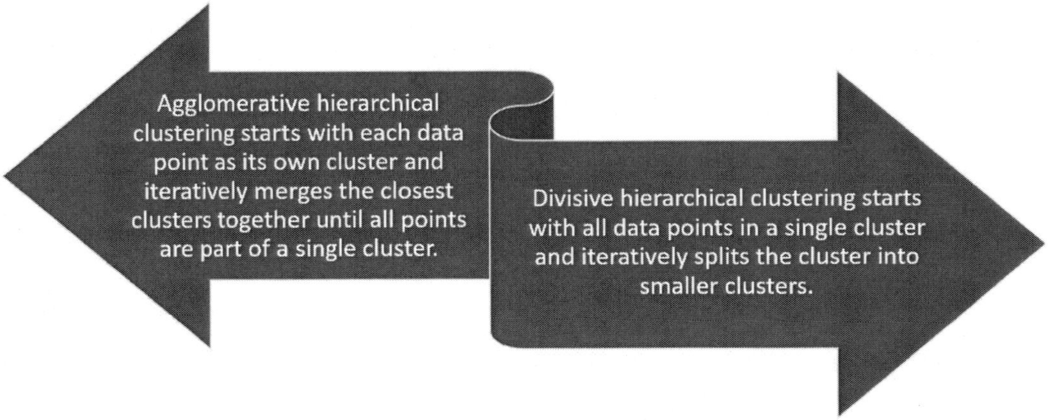

- Agglomerative hierarchical (bottom-up) clustering starts with each data point as its own cluster and iteratively merges the closest clusters together until all points are part of a single cluster.
- Divisive hierarchical (top-down) clustering starts with all data points in a single cluster and iteratively splits the cluster into smaller clusters.

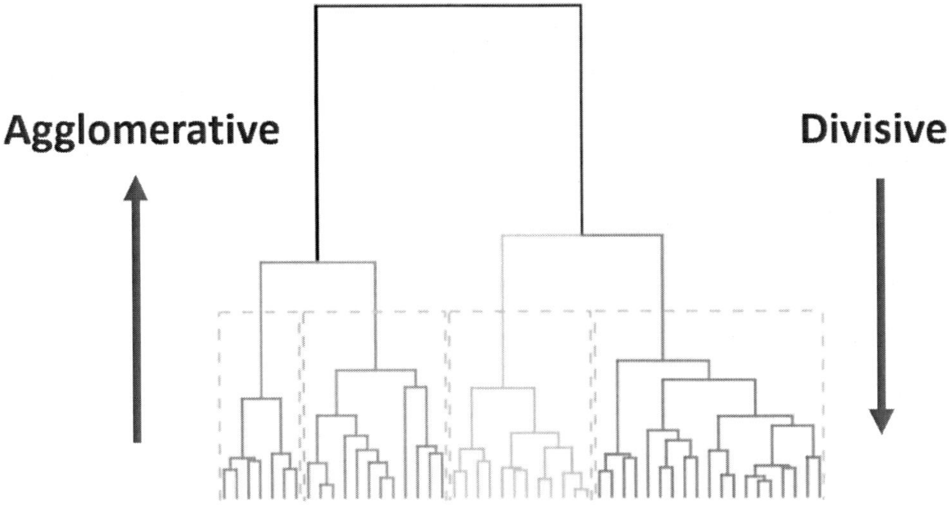

One of the main advantages of hierarchical clustering is that it allows for the representation of the hierarchy of clusters in a dendrogram, which can be used to visualize the relationships between different clusters. Another advantage is that it can handle non-linearly separable data and can be used with any distance metric.

However, one of the main disadvantages of hierarchical clustering is that it can be computationally expensive and time-consuming, especially for large datasets. Additionally, it can be sensitive to the choice of linkage criteria and distance metric used. Some commonly used linkage criteria include single linkage, complete linkage, average linkage, and Ward linkage.

An example of hierarchical clustering could be grouping customer data into segments for targeted marketing. By using clustering algorithms such as hierarchical clustering on customer data, such as their demographics, purchase history, and behavior, a company can create segments of similar customers that they can target with tailored marketing campaigns.

Hierarchical clustering is a method of clustering in which clusters are organized into a tree-like structure. It is also known as **hierarchical cluster analysis (HCA)** or agglomerative clustering.

Here is an example of how to perform agglomerative and divisive hierarchical clustering on a randomly generated dataset using Python and visualize the results with Matplotlib:

```
import numpy as np
import matplotlib.pyplot as plt
from scipy.cluster.hierarchy import dendrogram, linkage

# Generate random dataset with seed for reproducibility
np.random.seed(123)
X = np.random.randn(50, 2)
```

```
# Perform agglomerative clustering
Z_agg = linkage(X, method='ward')

# Plot dendrogram for agglomerative clustering
plt.figure(figsize=(10, 5))
plt.title("Agglomerative Clustering Dendrogram")
plt.xlabel("Data point index")
plt.ylabel("Distance")
dendrogram(Z_agg)
plt.show()

# Perform divisive clustering
Z_div = linkage(X.T, method='ward')

# Plot dendrogram for divisive clustering
plt.figure(figsize=(10, 5))
plt.title("Divisive Clustering Dendrogram")
plt.xlabel("Data point index")
plt.ylabel("Distance")
dendrogram(Z_div)
plt.show()
```

The output will look like this:

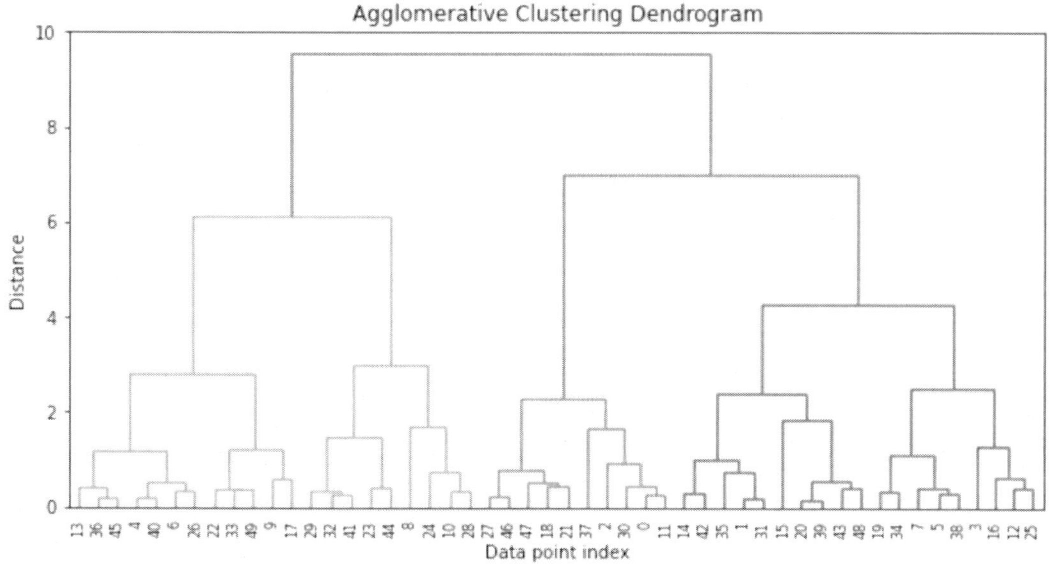

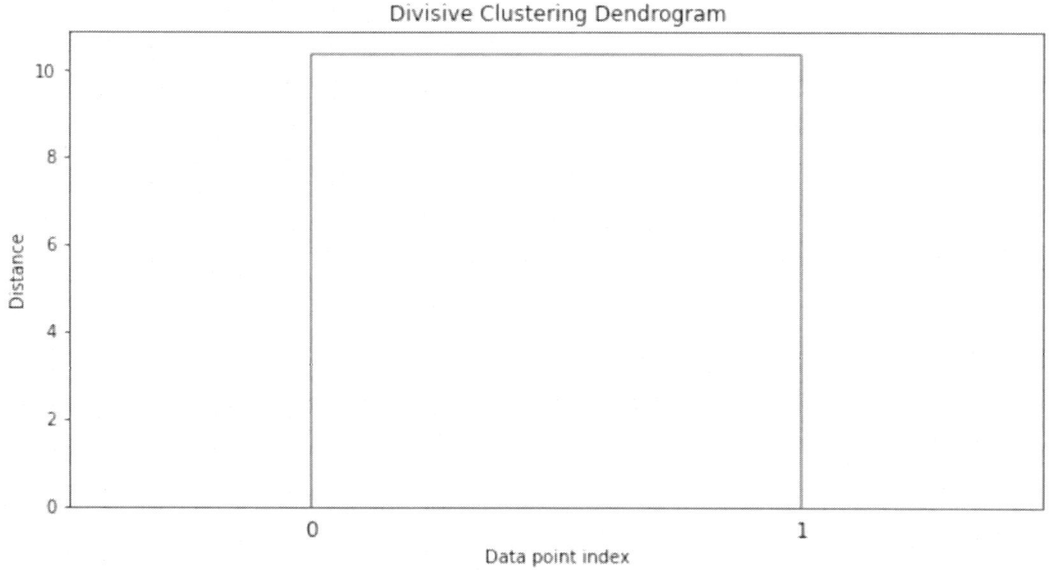

In the above example, we first generated a random dataset with 50 data points and 2 features using NumPy's **randn** function with seed 123. We then performed agglomerative hierarchical clustering using the **linkage** function from

scipy.cluster.hierarchy module with the **ward** method, which minimizes the variance of the distances between the clusters being merged. The resulting hierarchical structure is visualized with a dendrogram using Matplotlib's **dendrogram** function.

Next, we performed divisive hierarchical clustering by transposing the dataset and performing agglomerative clustering on it. This effectively treats the features as data points and the data points as features. The resulting hierarchical structure is also visualized with a dendrogram using Matplotlib.

Agglomerative clustering starts with each data point in its own cluster and merges the most similar clusters at each step until all data points belong to the same cluster. Divisive clustering starts with all data points in a single cluster and recursively divides the cluster into smaller and smaller clusters until each data point is in its own cluster.

The choice of linkage method can greatly affect the resulting clusters and dendrogram structures, and different linkage methods may be more appropriate for different datasets and clustering objectives. If we just change the linkage method in the code above, the result will look completely different:

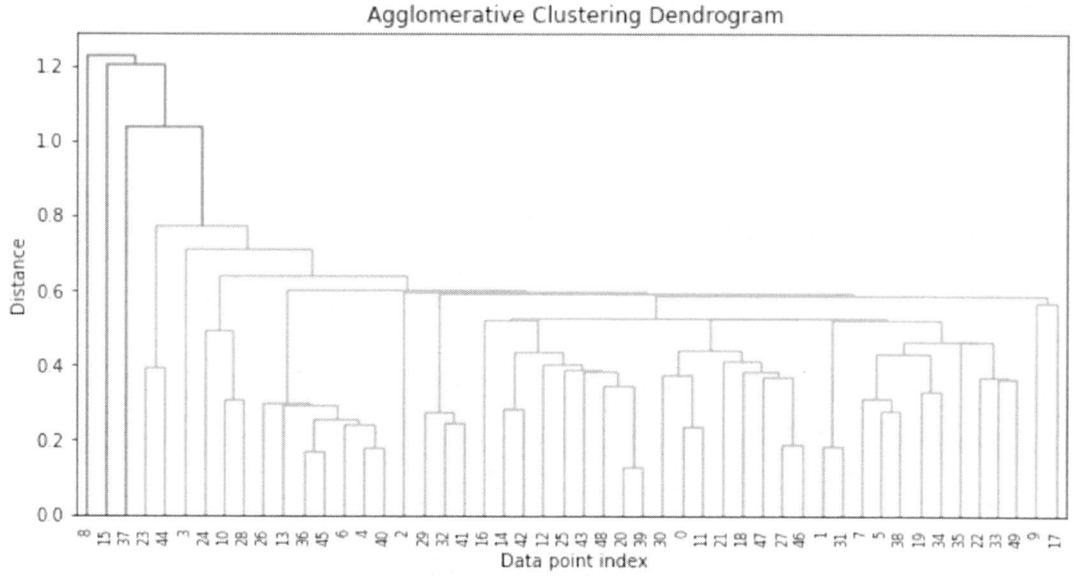

It's important to note that hierarchical clustering is a form of clustering that builds a hierarchy of clusters, with each node in the hierarchy representing a cluster. The leaves of the hierarchy are the individual data points, and each node is connected to its parent by a linkage criterion, such as single linkage, complete linkage, etc.

5.4 DBSCAN

DBSCAN (Density-Based Spatial Clustering of Applications with Noise) is a density-based clustering algorithm that is used to identify clusters of arbitrary shapes in large datasets. It is particularly useful when the data is not well-separated, and traditional clustering methods like K-means and hierarchical clustering may not work well.

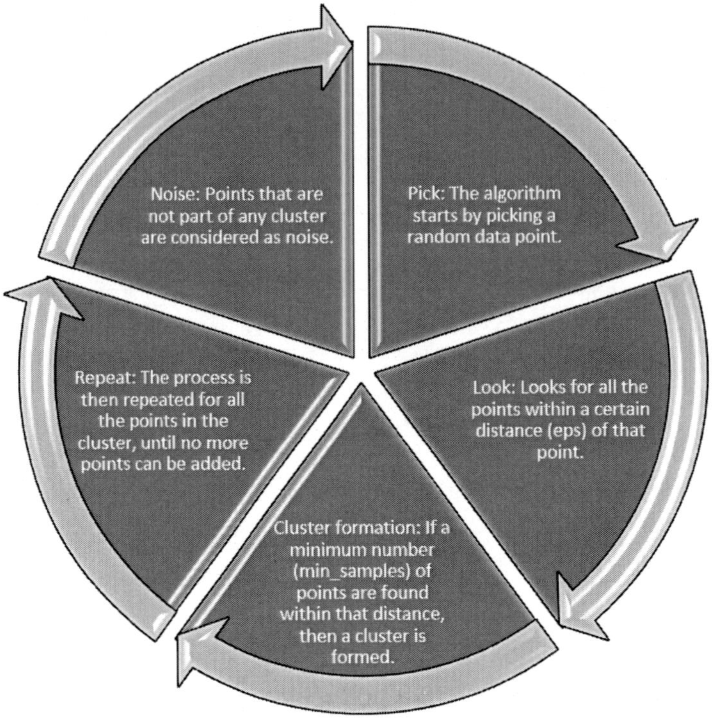

The basic idea behind DBSCAN is to group together data points that are close to each other in the feature space, based on a distance metric and a density threshold. The algorithm starts by picking a random data point, and then looks for all the points within a certain distance (eps) of that point. If a minimum number (min_samples) of points are found within that distance, then a cluster is formed. The process is then repeated for all the points in the cluster, until no more points can be added. Points that are not part of any cluster are considered as noise.

One of the key advantages of DBSCAN is that it can find clusters of **arbitrary shapes**, unlike K-means and hierarchical clustering that assume clusters to be spherical or hierarchical in shape. Additionally, DBSCAN does not require prior knowledge of the number of clusters in the data, as it automatically detects the number of clusters based on the density of the data.

However, one of the major disadvantages of DBSCAN is that it is sensitive to the choice of parameters, particularly eps and min_samples. Choosing the right values for these parameters can be difficult and requires some experimentation. Additionally, DBSCAN can be sensitive to the scale of the data, and may not work well when the data has varying densities or different types of features.

In Python, DBSCAN can be implemented using the scikit-learn library. The DBSCAN class in scikit-learn requires two main parameters: eps and min_samples. The eps parameter defines the maximum distance between two points for them to be considered as part of the same cluster, while the min_samples parameter defines the minimum number of points required to form a cluster. The algorithm can also take an optional metric parameter, which specifies the distance metric to use (default is Euclidean distance).

Example:

Here's a real-life example of how DBSCAN can be used to cluster customer data based on their purchasing behavior in a retail store.

Suppose a retail store has collected data on customers' purchasing behavior, including the amount spent, the number of items purchased, and the time spent in the store. The store wants to use this data to segment customers into different groups based on their purchasing behavior. DBSCAN can be used to identify groups of customers that have similar purchasing behavior.

Let's see the code below:

```
import numpy as np
from sklearn.cluster import DBSCAN
```

```
import matplotlib.pyplot as plt

# Generate random data
np.random.seed(0)
X = np.random.rand(100, 2)

# Add noise to the data
X[50:60, :] = 1.5 + 0.1 * np.random.randn(10, 2)

# Cluster the data using DBSCAN
db = DBSCAN(eps=0.3, min_samples=5).fit(X)

# Plot the results
labels = db.labels_
n_clusters_ = len(set(labels)) - (1 if -1 in labels else 0)
plt.figure(figsize=(8, 6))
unique_labels = set(labels)
colors = [plt.cm.Spectral(each) for each in np.linspace(0, 1,
len(unique_labels))]
for k, col in zip(unique_labels, colors):
    if k == -1:
        col = [0, 0, 0, 1]
    class_member_mask = (labels == k)
    xy = X[class_member_mask]
    plt.plot(xy[:, 0], xy[:, 1], 'o', markerfacecolor=tuple(col),
markersize=10)
plt.title('Estimated number of clusters: %d' % n_clusters_)
plt.show()
```

The output will look like this:

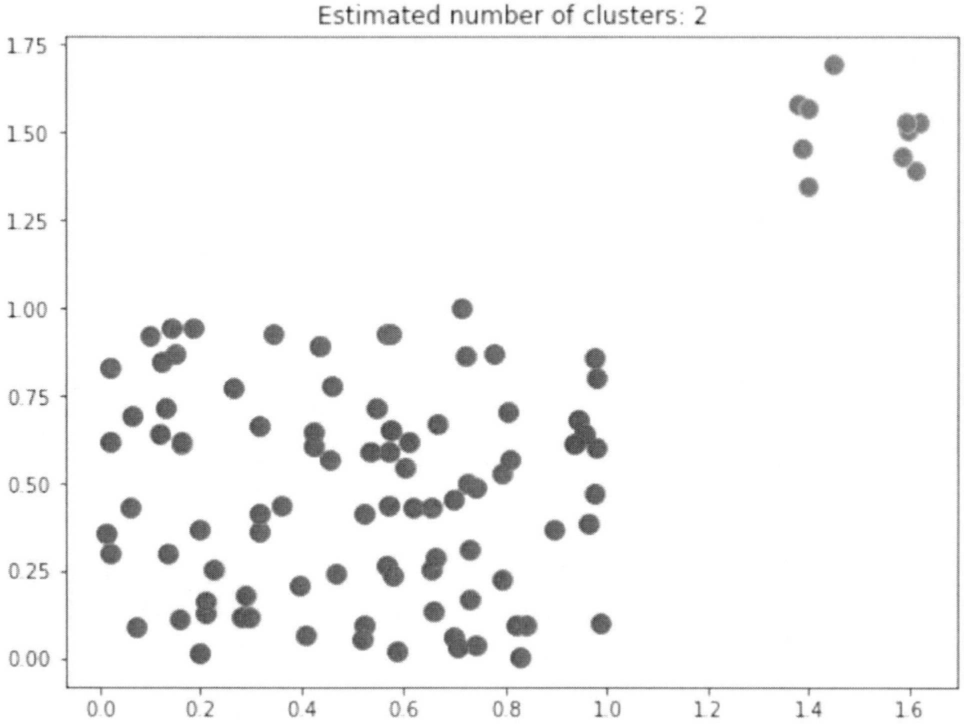

In this example, we first generate random data using the **numpy.random.rand** function. We then add some noise to the data by modifying a subset of the points. The **DBSCAN** algorithm is then used to cluster the data based on the **eps** and **min_samples** parameters. The results are plotted using **matplotlib.pyplot**.

The resulting plot shows the clustered data points, with each cluster assigned a unique color. The algorithm also identifies the noise points, which are shown in black.

This example shows how DBSCAN can be used to segment customers into different groups based on their purchasing behavior. The clusters can then be used to target specific groups of customers with tailored marketing campaigns or promotions.

5.5 GMM (GAUSSIAN MIXTURE MODEL)

Gaussian Mixture Model (GMM) is a generative probabilistic model that assumes all the data points are generated from a mixture of a finite number of Gaussian distributions with unknown parameters. It is a clustering algorithm that tries to find natural groupings in the data by assuming that the data points are generated from a mixture of Gaussian distributions with unknown parameters.

The GMM algorithm is a probabilistic model that assumes that the data points are generated from a mixture of Gaussian distributions. Each Gaussian distribution is represented by its mean, covariance, and weight. The weight represents the proportion of data points that belong to that Gaussian distribution. The GMM algorithm uses the Expectation-Maximization (EM) algorithm to estimate the parameters of the Gaussian distributions and the weights.

Step-by-Step Process

Here is a step-by-step process for the GMM algorithm:

1. Initialize the number of clusters K and the GMM model parameters: mean, covariance matrix, and mixing coefficients.
2. Expectation Step (E-step):
    a) Compute the probability density function (PDF) for each data point for each of the K Gaussian distributions.

    b) Compute the responsibility of each cluster for each data point, which is the conditional probability of the data point belonging to that cluster given the PDF value and the mixing coefficient.
3. Maximization Step (M-step):

a) Compute the updated mean, covariance matrix, and mixing coefficient for each cluster by using the responsibility values from the E-step.

b) The mixing coefficient represents the proportion of data points belonging to each cluster.

4. Calculate the log-likelihood of the data given the updated parameters.

5. Check the convergence criteria. If the log-likelihood is not changing significantly from one iteration to the next, stop the algorithm; otherwise, go back to step 2.

6. Return the cluster labels for each data point based on the maximum responsibility value.

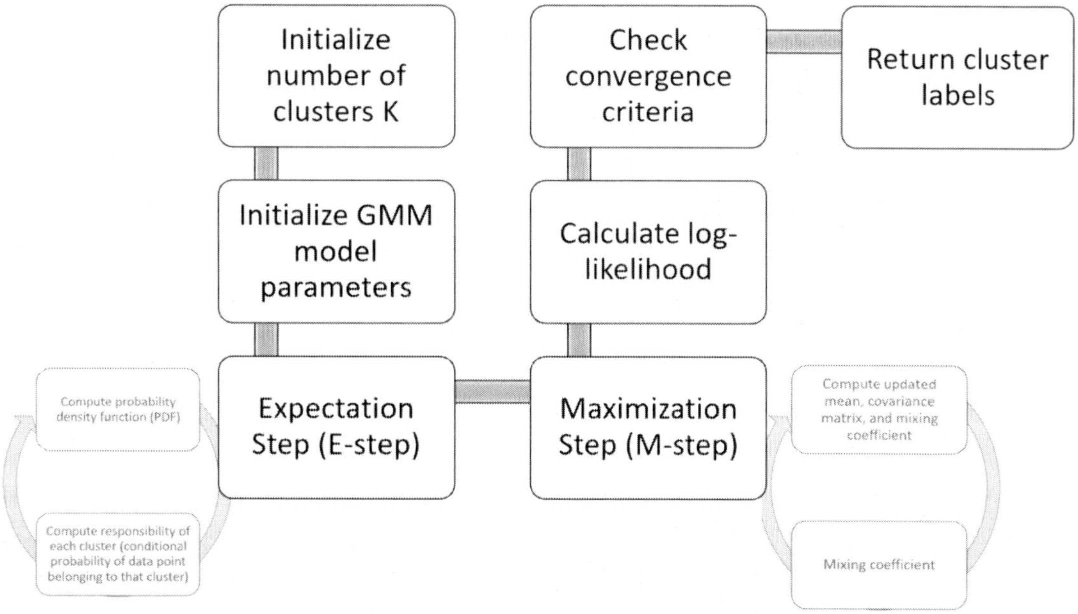

Note that steps 2 and 3 are iterated until convergence is achieved. The convergence criteria could be a maximum number of iterations, a small change in log-likelihood, or both. Additionally, the initialization of the model parameters could affect the final clustering result, so it is often useful to perform multiple runs of the algorithm with different initializations and choose the one with the highest log-likelihood.

The GMM algorithm is a soft clustering algorithm, which means that it assigns each data point a probability of belonging to each cluster. This is useful when the data points do not clearly belong to a single cluster or when the clusters have overlapping regions.

The GMM algorithm can be used for various types of data, including continuous data, categorical data, and mixed data. It is widely used in various applications such as image segmentation, speech recognition, and bioinformatics.

In Python, the GaussianMixture class of the sklearn.mixture module can be used to implement the GMM algorithm. The class takes the number of components (clusters) and the covariance type as input parameters. It also has various options to initialize the means, weights, and covariances of the Gaussian distributions. The fit method of the class is used to fit the GMM model to the data, and the predict method is used to predict the clusters for new data points.

Real-life use case: Customer Segmentation

In this example, we will use the Gaussian Mixture Model (GMM) algorithm to perform customer segmentation. Customer segmentation is a common use case in marketing, where the goal is to group customers based on their purchasing behavior, demographics, and other relevant features. These groups can be used to create targeted marketing campaigns, personalize customer experiences, and improve customer retention.

We will generate a random dataset that simulates customer purchasing behavior, where each data point represents a customer with features such as age, income, and purchase amount.

Code:

```
import numpy as np
import matplotlib.pyplot as plt
from sklearn.mixture import GaussianMixture

# Generate random dataset
np.random.seed(0)
n_samples = 500
X = np.zeros((n_samples, 2))
X[:200, :] = np.random.multivariate_normal(mean=[0, 0], cov=[[1, 0.3],
[0.3, 2]], size=200)
X[200:400, :] = np.random.multivariate_normal(mean=[5, 5], cov=[[1, -
0.3], [-0.3, 2]], size=200)
X[400:, :] = np.random.multivariate_normal(mean=[10, 0], cov=[[1, 0],
[0, 1]], size=100)

# Fit GMM model to data
gmm = GaussianMixture(n_components=3, random_state=0)
gmm.fit(X)

# Predict clusters for each data point
labels = gmm.predict(X)

# Plot data points colored by their predicted cluster
plt.scatter(X[:, 0], X[:, 1], c=labels)
plt.title('Customer Segmentation using GMM')
plt.xlabel('Age')
plt.ylabel('Purchase Amount')
plt.show()
```

The output will look like this:

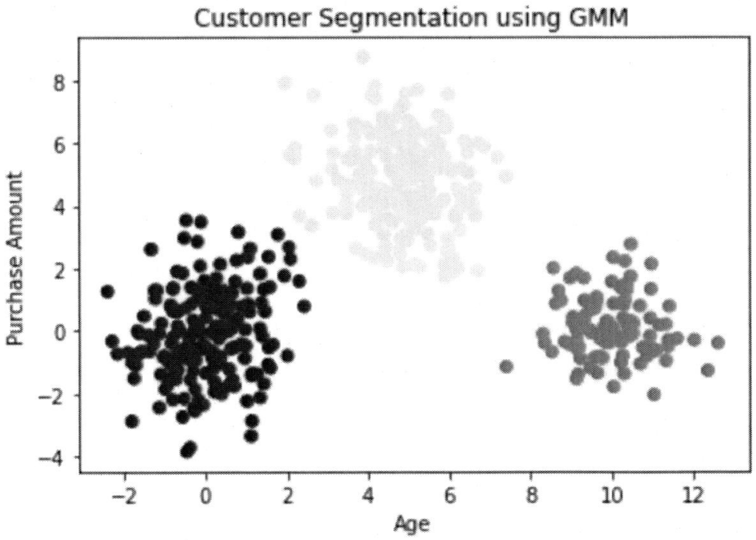

Explanation:

1. We first import the necessary libraries: numpy for generating random data, matplotlib for visualizing the results, and GaussianMixture from scikit-learn for fitting the GMM model.

2. We set the random seed to ensure reproducibility and generate a random dataset with 500 data points. We use the **multivariate_normal** function to generate data points from two different Gaussian distributions, and one set of points with a uniform distribution.

3. We initialize a GMM object with 3 components and fit it to the generated data using the **fit** method.

4. We predict the cluster labels for each data point using the **predict** method.

5. Finally, we plot the data points colored by their predicted cluster using **scatter** method of matplotlib.

The resulting plot shows three distinct customer segments based on their purchasing behavior: younger customers with lower purchase amounts (cluster 0), older customers with higher purchase amounts (cluster 1), and middle-aged customers with moderate purchase amounts (cluster 2).

GMM algorithm can be applied to many other use cases such as image segmentation, anomaly detection, and speech recognition.

5.6 DIMENSIONALITY REDUCTION

Dimensionality reduction is a technique used to reduce the number of features in a dataset while retaining as much information as possible. The goal of dimensionality reduction is to reduce the complexity of the data and make it more manageable for analysis and modeling. There are many different techniques used for dimensionality reduction, including principal component analysis (PCA), linear discriminant analysis (LDA), and t-distributed stochastic neighbor embedding (t-SNE). These techniques can be used individually or in combination to achieve the desired level of dimensionality reduction.

PCA is one of the most commonly used dimensionality reduction techniques. It works by transforming the original features into a new set of features, called principal components, which are linear combinations of the original features. The principal components are chosen such that they explain the most variance in the data. LDA is similar to PCA, but it is used for supervised learning and is used to find the linear combinations of features that best separate different classes.

t-SNE is a non-linear dimensionality reduction technique that is particularly useful for visualizing high-dimensional data in two or three dimensions. It works by constructing a probability distribution over the high-dimensional data and then mapping it to a lower-dimensional space while preserving the structure of the data as much as possible.

Overall, dimensionality reduction techniques are important tools for data analysis and modeling, as they can help to simplify the data and make it more manageable for analysis and modeling. It also helps in better visualization and interpreting the data.

Let's discuss some of dimensionality reduction techniques in subsequent sections.

5.7 PRINCIPAL COMPONENT ANALYSIS (PCA)

Principal Component Analysis (PCA) is a popular dimensionality reduction technique used to transform a high-dimensional dataset into a lower-dimensional space while retaining as much of the original information as possible. The main idea behind PCA is to identify the directions of maximum variance in the data and project the data onto a new coordinate system defined by these directions.

Step-by-Step Process

The following are the step-by-step processes involved in the PCA algorithm:

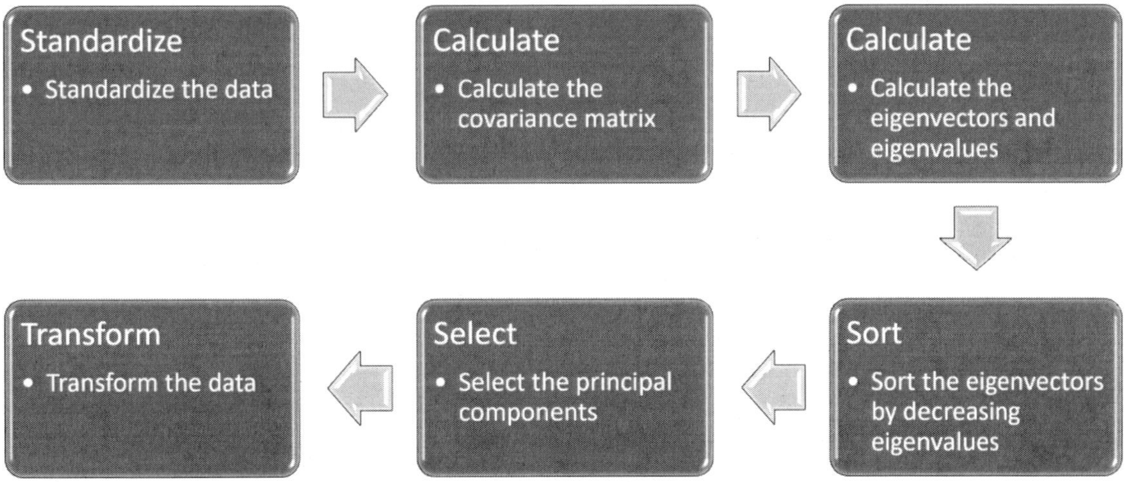

1. **Standardize the data**: The first step in PCA is to standardize the data to have a mean of 0 and a standard deviation of 1. This is done to ensure that all variables contribute equally to the analysis.

2. **Calculate the covariance matrix**: The next step is to calculate the covariance matrix of the standardized data. The covariance matrix shows the relationships between the variables in the dataset. The diagonal elements of the matrix represent the variances of the variables, while the off-diagonal elements represent the covariances between the variables.

3. **Calculate the eigenvectors and eigenvalues**: The next step is to calculate the eigenvectors and eigenvalues of the covariance matrix. Eigenvectors are a set of vectors that define the directions of the new feature space, while eigenvalues represent the magnitude of the variance of the data in the corresponding eigenvector direction.

4. **Sort the eigenvectors by decreasing eigenvalues**: The eigenvectors are sorted in descending order of their corresponding eigenvalues. The eigenvector with the highest eigenvalue represents the direction of

maximum variance in the data, while the eigenvector with the lowest eigenvalue represents the direction of minimum variance.

5. **Select the principal components**: The next step is to select the principal components based on the number of dimensions in the new feature space. The number of principal components selected should be less than or equal to the number of original dimensions in the dataset.

6. **Transform the data**: Finally, the data is transformed into the new feature space by multiplying the original data by the selected principal components. The transformed data can then be used for further analysis or visualization.

PCA is a powerful tool for dimensionality reduction, visualization, and data compression. It is widely used in various fields such as finance, image processing, genetics, and neuroscience, among others. One of the main advantages of PCA is that it can be used to remove noise from the data by eliminating the directions of minimum variance, thus improving the performance of machine learning models.

Example

Suppose we have a dataset containing information about customers' purchases at a grocery store. The dataset includes the price, quantity, and category of each item purchased, as well as the customer's age, gender, and location. We want to analyze this dataset to gain insights into customer behavior, but the dataset has a large number of features (i.e., columns), making it difficult to analyze.

To use PCA to reduce the dimensionality of this dataset, we can follow these steps:

1. Generate a random dataset with 100 samples, where each sample has 10 features using numpy.

2. Scale the dataset so that each feature has zero mean and unit variance using StandardScaler from scikit-learn.

3. Fit the PCA model to the scaled dataset using PCA from scikit-learn. We can specify the number of components we want to keep in the transformed dataset.

4. Transform the dataset using the fitted PCA model. The transformed dataset will have the specified number of components, which are linear combinations of the original features.

5. Visualize the transformed dataset using a scatter plot, where each point represents a sample in the transformed dataset. We can use different colors to represent different labels if we have them in the dataset.

Here's the complete code:

```
import numpy as np
import matplotlib.pyplot as plt
from sklearn.decomposition import PCA
from sklearn.preprocessing import StandardScaler

# Step 1: Generate a random dataset
np.random.seed(42)
X = np.random.rand(100, 10)

# Step 2: Scale the dataset
scaler = StandardScaler()
X_scaled = scaler.fit_transform(X)

# Step 3: Fit the PCA model
n_components = 2
pca = PCA(n_components=n_components)
pca.fit(X_scaled)

# Step 4: Transform the dataset
X_transformed = pca.transform(X_scaled)

# Step 5: Visualize the transformed dataset
```

```
plt.scatter(X_transformed[:, 0], X_transformed[:, 1])
plt.xlabel('PC1')
plt.ylabel('PC2')
plt.title('Transformed dataset')
plt.show()
```

The output will look like this:

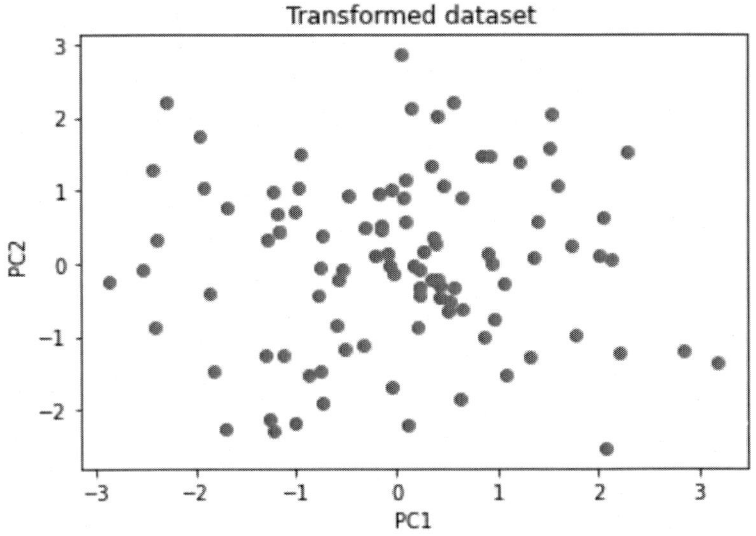

In this example, we generate a random dataset with 100 samples and 10 features using numpy. We then scale the dataset using StandardScaler from scikit-learn to ensure that each feature has zero mean and unit variance. We fit the PCA model to the scaled dataset and specify that we want to keep the first two components in the transformed dataset. We then transform the dataset using the fitted PCA model and plot the transformed dataset using a scatter plot.

The resulting plot shows the transformed dataset in two dimensions, where each point represents a sample in the transformed dataset. We can see that the samples are now more spread out along the two principal components than they were in the original dataset, which had 10 features. This can help us gain insights into the underlying structure of the data and potentially make it easier to analyze.

5.8 INDEPENDENT COMPONENT ANALYSIS (ICA)

Independent Component Analysis (ICA) is a technique used for separating a multivariate signal into independent, non-Gaussian components. It is a popular technique in the field of signal processing and has been applied in various fields such as audio, image, and bio-medical signal processing.

The main idea behind ICA is to find a linear combination of the original variables such that the resulting components are statistically independent. In other words, the goal is to find a new set of variables that are as independent as possible from each other. This can be achieved by maximizing the non-Gaussianity of the resulting components.

The ICA algorithm is typically applied to data that has been preprocessed with some form of whitening or decorrelation technique. This is because ICA is sensitive to the correlation structure of the data, and decorrelating the data beforehand can help improve the performance of the algorithm.

One common method for implementing ICA is called FastICA. This algorithm is based on the concept of negentropy and uses an iterative optimization method to find the independent components. Other popular methods include JADE and Infomax.

Step-by-Step Process

Here are the general steps for Independent Component Analysis (ICA):

1. Data Preprocessing: Preprocess the data by centering the data (subtracting the mean) and scaling it (dividing by the standard deviation or range).

2. **Whitening the Data**: Apply a linear transformation to the preprocessed data to transform it to a new set of uncorrelated variables with a diagonal covariance matrix (whitening). The diagonal elements of the covariance matrix represent the variance of each variable.

3. **Choosing the Number of Components**: Determine the number of independent components to extract. This can be done by analyzing the scree plot or by setting a threshold on the eigenvalues of the covariance matrix.

4. **Computing the Components**: Use an iterative algorithm (such as FastICA) to extract the independent components. The algorithm maximizes the non-Gaussianity of the transformed data in each dimension.

5. **Reconstructing the Data**: Reconstruct the original data from the extracted independent components by multiplying the components by the mixing matrix and adding the mean.

6. **Interpretation of Results**: Interpret the independent components in the context of the application.

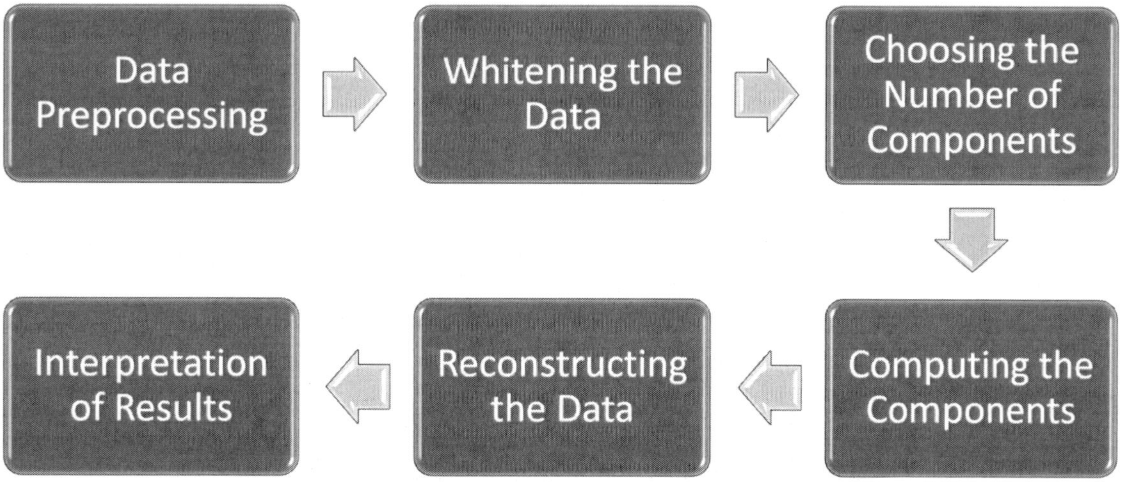

In Python, ICA can be implemented using the scikit-learn library. The library provides a class called FastICA, which can be used to fit an ICA model to the data. The class takes several parameters such as the number of components and the algorithm to use. Once the model is fitted, it can be used to transform the original data into the independent components.

Real-world applications of ICA include audio source separation, signal denoising, and feature extraction in bio-medical signals. For example, in audio source separation, ICA can be used to separate different sources of sound in a recording, such as vocals and instruments. In bio-medical signal processing, ICA can be used to extract independent components from EEG signals, which can then be used for diagnosis and treatment of neurological disorders.

Independent Component Analysis (ICA) is a technique used for identifying independent sources within a signal, which can then be separated from one another. In ICA, the goal is to find a linear combination of the input features that maximizes the non-Gaussianity of the resulting signals.

Example

Here is a coding example to illustrate Independent Component Analysis (ICA) using a random dataset:

```
import numpy as np
from scipy import signal
from sklearn.decomposition import FastICA
import matplotlib.pyplot as plt

# set random seed for reproducibility
np.random.seed(42)

# generate random mixed signals
n_samples = 2000
time = np.linspace(0, 8, n_samples)
s1 = np.sin(2 * time)  # signal 1
s2 = np.sign(np.sin(3 * time))  # signal 2
s3 = signal.sawtooth(2 * np.pi * time)  # signal 3
S = np.c_[s1, s2, s3]
S += 0.2 * np.random.normal(size=S.shape)  # add noise
S /= S.std(axis=0)  # standardize the data

# mix signals randomly
A = np.array([[0.5, 1, 1], [1, 0.5, 1], [1, 1, 0.5]])  # mixing matrix
X = np.dot(S, A.T)  # mixed signals

# apply Independent Component Analysis (ICA)
ica = FastICA(n_components=3)
S_hat = ica.fit_transform(X)

# plot the original and recovered signals
fig, axes = plt.subplots(3, sharex=True, figsize=(8, 8))
ax1, ax2, ax3 = axes

ax1.plot(S[:, 0], color='r')
ax1.set_title('Original Signal 1')

ax2.plot(S[:, 1], color='g')
ax2.set_title('Original Signal 2')

ax3.plot(S[:, 2], color='b')
ax3.set_title('Original Signal 3')

fig2, axes2 = plt.subplots(3, sharex=True, figsize=(8, 8))
```

```
ax4, ax5, ax6 = axes2

ax4.plot(S_hat[:, 0], color='r')
ax4.set_title('Recovered Signal 1')

ax5.plot(S_hat[:, 1], color='g')
ax5.set_title('Recovered Signal 2')

ax6.plot(S_hat[:, 2], color='b')
ax6.set_title('Recovered Signal 3')

plt.show()
```

The output will look like his:

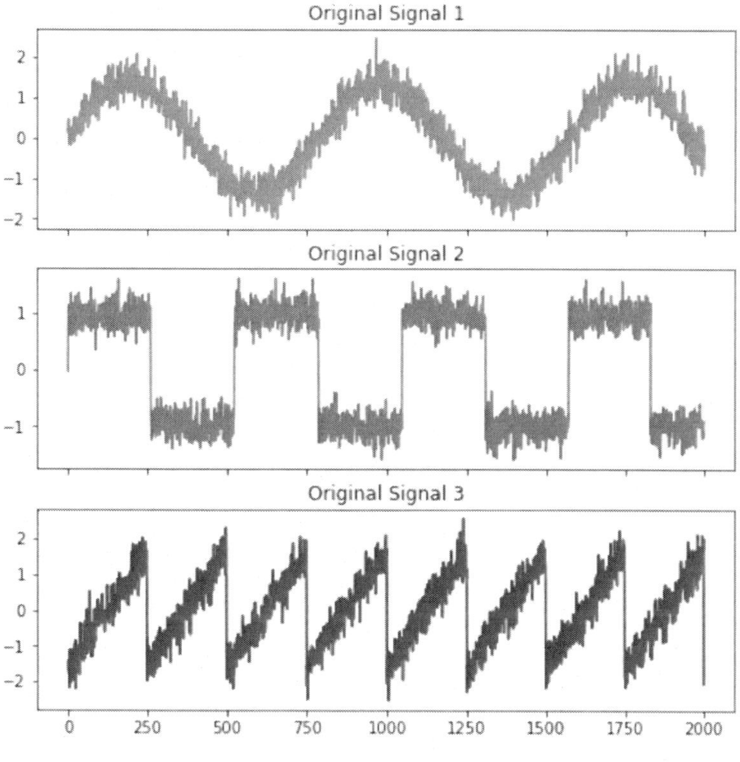

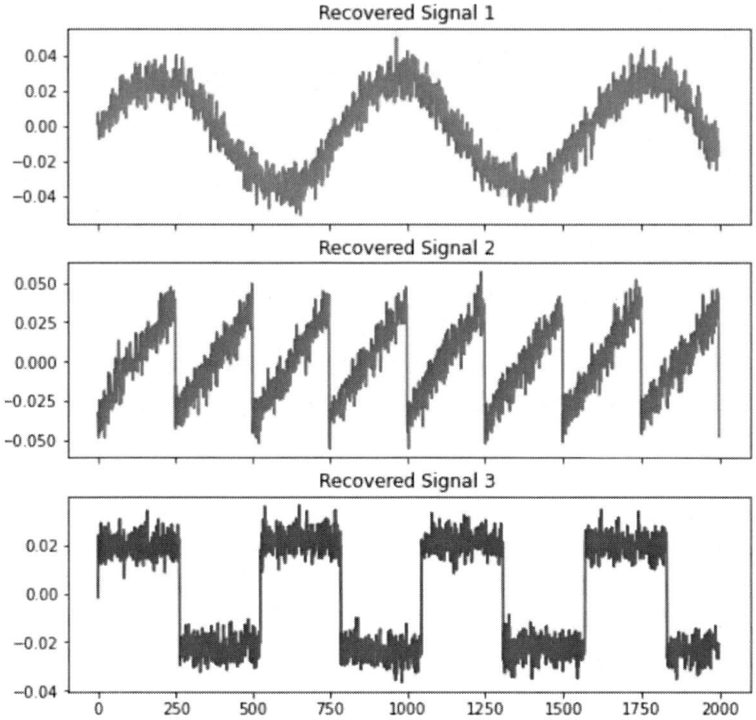

Explanation:

1. We first import the necessary packages: NumPy for generating random data, SciPy for signal processing, scikit-learn for implementing FastICA, and Matplotlib for visualizing the results.

2. We set the random seed for reproducibility.

3. We generate three random signals with different patterns (sine, square, and sawtooth waves), add some Gaussian noise to them, and standardize the data.

4. We mix the signals randomly by multiplying them with a mixing matrix.

5. We apply the FastICA algorithm to the mixed signals to extract the independent components.

6. We plot the original and recovered signals for comparison.

In this example, we use ICA to separate the mixed signals into their original components. This technique can be useful in various fields such as biomedical signal processing, speech recognition, and image processing.

5.9 T-SNE

t-Distributed Stochastic Neighbor Embedding (t-SNE) is a dimensionality reduction technique that is particularly well-suited for visualizing high-dimensional data. It is a non-linear dimensionality reduction technique that is based on probability distributions with random walk on neighborhood graphs to find structure in the data.

t-SNE works by minimizing the divergence between two probability distributions: a distribution that measures pairwise similarities between the datapoints in the high-dimensional space, and a distribution that measures pairwise similarities between the datapoints in the low-dimensional space (i.e. the space where we want to represent the data). The t-SNE algorithm maps the high-dimensional data to a low-dimensional space while preserving the structure of the data, such as clusters or patterns.

The t-SNE algorithm has two main components: the similarity matrix and the low-dimensional embedding. The similarity matrix is computed using a similarity metric such as the Euclidean distance or the cosine similarity. The low-dimensional embedding is computed using gradient descent. The t-SNE algorithm iteratively updates the low-dimensional embedding to minimize the divergence between the two probability distributions.

Steep-by-Step Process

The following are the steps involved in t-SNE:

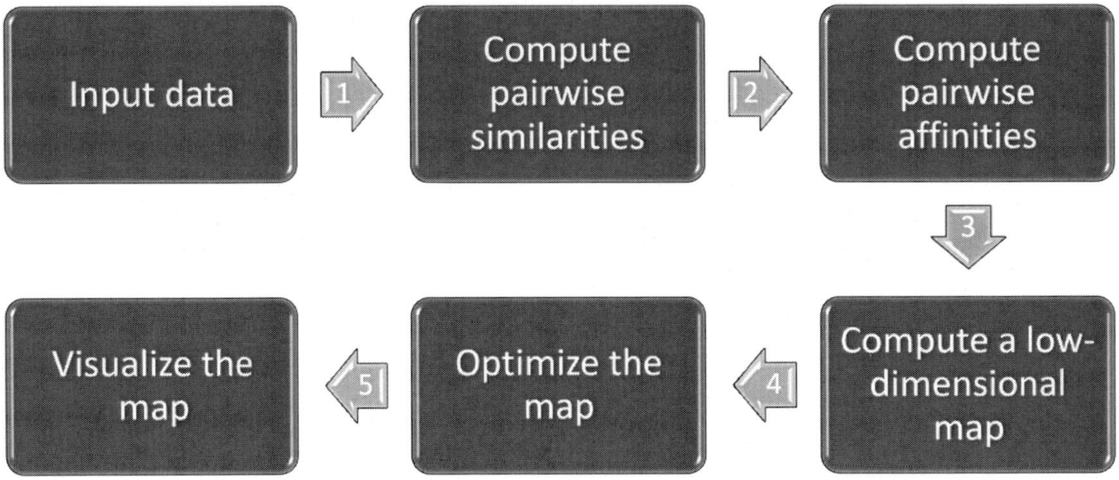

1. **Input data**: The first step in t-SNE is to provide the input data, which is usually a high-dimensional dataset with a large number of features.

2. **Compute pairwise similarities**: t-SNE algorithm calculates the pairwise similarity between all the data points using a Gaussian distribution. The Gaussian distribution measures the probability that two points are similar.

3. **Compute pairwise affinities**: The algorithm converts the pairwise similarities into a joint probability distribution over pairs of data points in such a way that similar points have a high probability of being picked as neighbors, while dissimilar points have a low probability of being picked.

4. **Compute a low-dimensional map**: t-SNE algorithm constructs a low-dimensional map by minimizing the divergence between two probability distributions: the joint probability distribution over pairs of data points in

the high-dimensional space and the distribution over pairs of corresponding points in the low-dimensional map.

5. **Optimize the map**: t-SNE algorithm optimizes the map using gradient descent to minimize the Kullback-Leibler divergence between the two probability distributions. The gradient descent is performed iteratively, and at each iteration, the algorithm updates the positions of the points in the low-dimensional space.
6. **Visualize the map**: The final step is to visualize the low-dimensional map using a scatter plot. The scatter plot shows the positions of the points in the low-dimensional space.

t-SNE is particularly useful for visualizing high-dimensional data, such as data with many features or large datasets. It has been used in a variety of applications, such as natural language processing, computer vision, and bioinformatics. However, t-SNE is sensitive to the choice of parameters and the initialization of the low-dimensional embedding, and can be computationally expensive for large datasets.

Example

In this example, we will generate a random dataset of 1000 2-dimensional points and then use t-SNE to visualize the dataset in a lower-dimensional space (2D space).

```
import numpy as np
from sklearn.manifold import TSNE
import matplotlib.pyplot as plt

# Generate random dataset
np.random.seed(42)
X = np.random.rand(1000, 2)
```

```
# Apply t-SNE to reduce dimensionality
tsne = TSNE(n_components=2, perplexity=30.0)
X_tsne = tsne.fit_transform(X)

# Visualize the dataset in 2D space
plt.scatter(X_tsne[:, 0], X_tsne[:, 1])
plt.title("t-SNE visualization of random dataset")
plt.show()
```

The output will look like this:

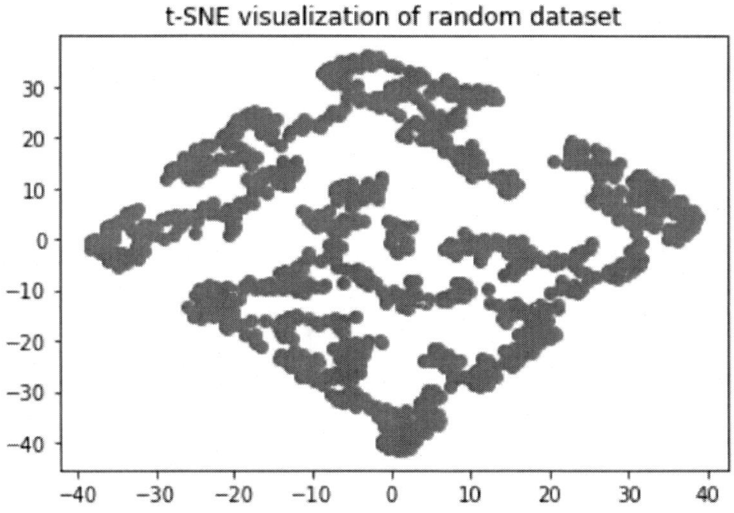

Explanation:

1. We first import the necessary libraries: NumPy for generating a random dataset, t-SNE from scikit-learn for applying t-SNE, and Matplotlib for visualizing the dataset.

2. We set a random seed for reproducibility and generate a random dataset of 1000 2-dimensional points using the **np.random.rand()** function.

3. We create an instance of the **TSNE** class and set the number of components to 2 (since we want to visualize the dataset in 2D space) and the perplexity to 30.0 (a hyperparameter that controls the balance between preserving global and local structure of the dataset).

4. We apply t-SNE to the dataset by calling the **fit_transform()** method of the **TSNE** instance.

5. Finally, we plot the 2D t-SNE embeddings using Matplotlib's **scatter()** function and give a title to the plot.

The output plot will show a scatter plot of the 1000 points in 2D space, where the points are arranged in a way that preserves the underlying structure of the original dataset.

5.10 AUTOENCODERS

Autoencoders are a type of neural network architecture that are designed to learn a compressed representation of input data. They consist of two main components: an encoder, which maps the input data to a lower-dimensional representation, and a decoder, which maps the lower-dimensional representation back to the original input data. The goal of the autoencoder is to learn a compressed representation of the input data that captures the most important features or patterns.

One common use case for autoencoders is dimensionality reduction. By training an autoencoder to learn a lower-dimensional representation of the input data, it can be used to reduce the number of features in a dataset while still preserving the most important information. Autoencoders can also be used for anomaly detection, by training the model on normal data and using it to identify data points that are significantly different from the training data.

Autoencoders can be implemented using various types of neural network architectures, such as feedforward neural networks, recurrent neural networks, and convolutional neural networks. The choice of architecture will depend on the specific problem and the type of input data.

Step-by-Step Process

Here is a step-by-step process for implementing autoencoders:

1. **Gather and preprocess data**: As with any machine learning task, start by gathering your data and preprocess it as necessary (e.g. normalize, standardize, feature selection).

2. **Choose an architecture**: Autoencoders have a specific architecture consisting of an encoder and a decoder. The encoder reduces the input data to a lower-dimensional representation, while the decoder then reconstructs the original data from this lower-dimensional representation. Decide on the number of layers and neurons in each layer for both the encoder and decoder.

3. **Define the loss function**: In order to train the autoencoder, you need to define a loss function that measures the difference between the original input data and the reconstructed output data. Mean squared error is a common choice.

4. **Train the autoencoder**: Train the autoencoder using your chosen optimization algorithm and the defined loss function. Use your preprocessed data as input and output.

5. **Evaluate performance**: After training, evaluate the performance of your autoencoder using metrics like mean squared error or visual inspection of the reconstructed data.

6. **Use the encoder for dimensionality reduction**: The encoder part of the trained autoencoder can be used as a dimensionality reduction technique for new data that was not seen during training.

7. **Fine-tune the autoencoder**: If necessary, fine-tune the autoencoder by tweaking hyperparameters and re-training.

8. **Deploy the autoencoder**: Once you are satisfied with the performance of the autoencoder, deploy it in your production environment.

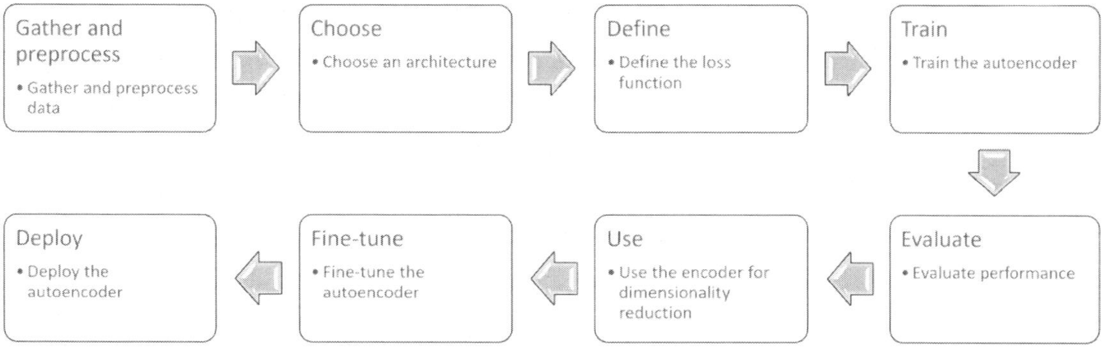

Example

Here's an example of implementing an autoencoder using a randomly generated dataset:

```
import numpy as np
import matplotlib.pyplot as plt
from tensorflow import keras

# Generate random dataset
np.random.seed(42)
n_samples = 1000
input_dim = 10
X = np.random.rand(n_samples, input_dim)

# Define autoencoder model
input_layer = keras.layers.Input(shape=(input_dim,))
encoded = keras.layers.Dense(5, activation='relu')(input_layer)
decoded = keras.layers.Dense(input_dim, activation='sigmoid')(encoded)
autoencoder = keras.models.Model(input_layer, decoded)

# Compile and train model
autoencoder.compile(optimizer='adam', loss='binary_crossentropy')
autoencoder.fit(X, X, epochs=50, batch_size=32)

# Use the encoder part of the autoencoder to get the encoded
representation of the input data
encoder = keras.models.Model(input_layer, encoded)
encoded_X = encoder.predict(X)

# Plot the original and encoded data
fig, axs = plt.subplots(nrows=1, ncols=2, figsize=(10, 5))
axs[0].scatter(X[:, 0], X[:, 1], c='b', alpha=0.5)
axs[0].set_title('Original Data')
axs[1].scatter(encoded_X[:, 0], encoded_X[:, 1], c='r', alpha=0.5)
axs[1].set_title('Encoded Data')
plt.show()
```

The output will look like this:

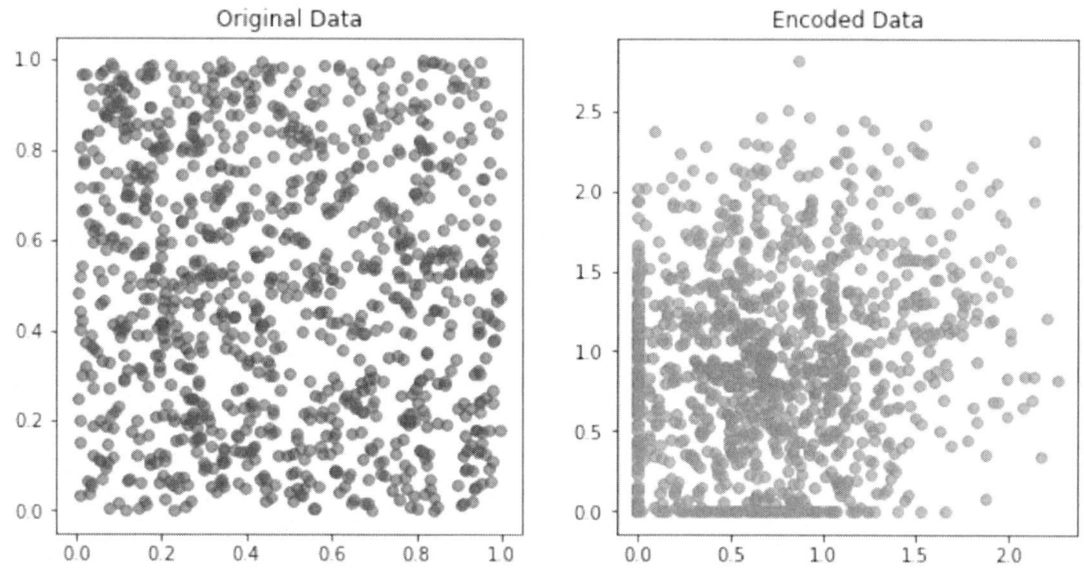

In this example, we generate a random dataset with 1000 samples of 10 dimensions. We then define an autoencoder model using Keras, with an input layer of size 10, an encoding layer of size 5, and a decoding layer of size 10. We compile the model using the binary crossentropy loss function and the Adam optimizer, and train it for 50 epochs using batches of 32 samples.

After training, we use the encoder part of the autoencoder to get the encoded representation of the input data, and plot the original and encoded data using matplotlib. The plot shows that the autoencoder has learned a lower-dimensional representation of the original data, with only 5 dimensions instead of 10, while preserving important features of the data.

5.11 ANOMALY DETECTION

Anomaly detection, also known as outlier detection or novelty detection, is the process of identifying data points that do not conform to the expected pattern or normal behavior of a given dataset. These data points, also known as anomalies or outliers, can be caused by errors in data collection or measurement, or they can represent truly abnormal or rare events.

There are various techniques for anomaly detection, including statistical methods, machine learning algorithms, and domain-specific methods. Some common statistical methods include using the mean and standard deviation to identify data points that fall outside of a certain range, or using the Z-score to identify data points that are a certain number of standard deviations away from the mean.

Machine learning algorithms such as clustering and density-based methods can also be used for anomaly detection. Clustering algorithms group similar data points together and identify data points that do not belong to any cluster as anomalies. Density-based methods, on the other hand, identify data points that are in low-density regions of the dataset as anomalies.

There are also domain-specific methods for anomaly detection that are tailored to specific types of data and applications. For example, in the field of cyber security, intrusion detection systems use various techniques to identify abnormal behavior in network traffic. In the field of finance, fraud detection systems use various techniques to identify abnormal transactions.

Anomaly detection is an important task in many fields, including finance, healthcare, cybersecurity, and manufacturing. It can be used to detect fraud, identify equipment failures, detect intrusions, and more. However, it is important to note that it is not always easy to determine whether a data point is truly an anomaly or not, and it can be difficult to distinguish between normal variations and truly abnormal events.

Step-by-Steep Process

The step-by-step process for anomaly detection can be as follows:

1. **Define the problem and identify the data**: First, we need to define the problem we are trying to solve and identify the relevant data sources. For example, we may want to detect anomalies in user behavior on a website, in sensor data from a machine, or in financial transaction data.

2. **Preprocess the data**: Once we have identified the data, we need to preprocess it to make it suitable for analysis. This may include steps such as cleaning the data, transforming it into a suitable format, and normalizing it to ensure that all features are on the same scale.

3. **Select an appropriate algorithm**: There are several different algorithms that can be used for anomaly detection, including statistical methods, machine learning models, and deep learning approaches. The choice of algorithm will depend on the nature of the data and the specific requirements of the problem.

4. **Train the model**: Once we have selected an appropriate algorithm, we need to train the model on the available data. This may involve splitting the data into training and validation sets, selecting appropriate hyperparameters for the model, and tuning the model to achieve the best possible performance.

5. **Evaluate the model**: After training the model, we need to evaluate its performance on a separate test set. This will allow us to estimate how well the model will perform on new, unseen data.

6. **Deploy the model**: Once we are satisfied with the performance of the model, we can deploy it in a production environment to detect anomalies in real-time. This may involve integrating the model with other systems, setting appropriate thresholds for anomaly detection, and monitoring the performance of the model over time.

7. **Iterate and improve**: Anomaly detection is an ongoing process, and it is important to continually monitor the performance of the model and make improvements as necessary. This may involve collecting additional data, retraining the model on new data, or fine-tuning the model to improve its performance.

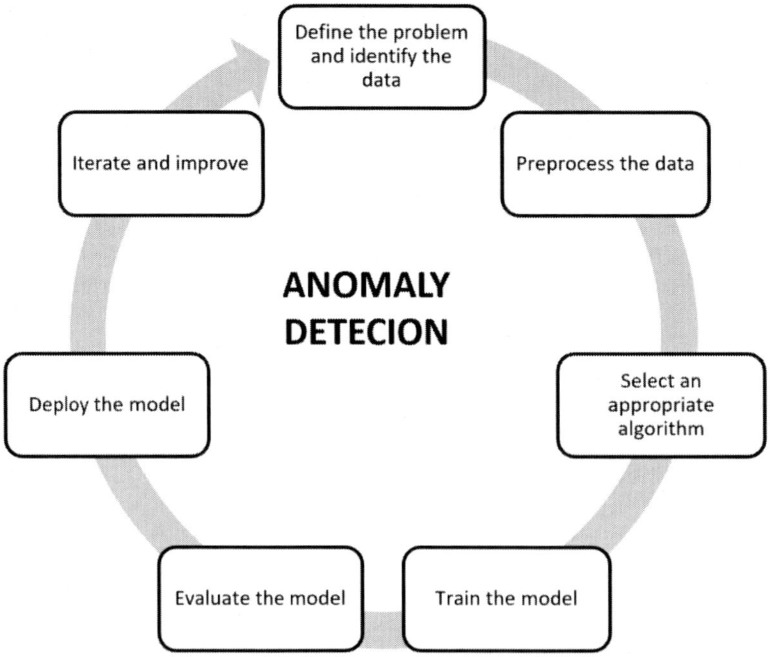

Example (using the Isolation Forest algorithm)

here is an example of anomaly detection using the Isolation Forest algorithm on a randomly generated dataset:

```
import numpy as np
import matplotlib.pyplot as plt
from sklearn.ensemble import IsolationForest

# Generate random dataset with anomalies
rng = np.random.RandomState(42)
X = 0.3 * rng.randn(100, 2)
X_train = np.r_[X + 2, X - 2, np.random.uniform(low=-4, high=4,
size=(20, 2))]

# Fit the isolation forest model
clf = IsolationForest(random_state=rng, contamination=0.1)
clf.fit(X_train)

# Predict anomalies
y_pred = clf.predict(X_train)
outliers = X_train[np.where(y_pred == -1)]

# Plot the dataset and anomalies
plt.scatter(X_train[:, 0], X_train[:, 1], c='white', edgecolor='k',
alpha=0.5)
plt.scatter(outliers[:, 0], outliers[:, 1], c='red', edgecolor='k')
plt.title('Isolation Forest for Anomaly Detection')
plt.show()
```

The output will look like this:

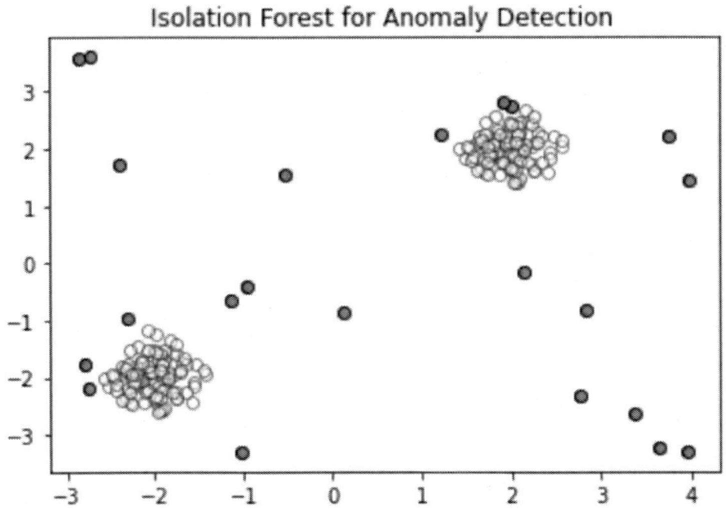

In this example, we first generate a random dataset with 100 normal samples and 20 anomalous samples. We then fit an Isolation Forest model to the dataset with a contamination rate of 0.1, which specifies the proportion of anomalies we expect to have in the dataset. The model is then used to predict the anomalies, and we plot the dataset and the predicted anomalies with red dots.

The Isolation Forest algorithm works by constructing isolation trees for the dataset, which are binary trees that randomly select a feature and split the data along a randomly selected value for that feature. Anomalies are typically found to have shorter path lengths from the root node to the leaf nodes, since they are isolated more easily than normal points. The Isolation Forest algorithm takes advantage of this property to detect anomalies.

Example (using the Gaussian Distribution)

One popular method of anomaly detection is using the Gaussian Distribution. Here's an example of how to implement anomaly detection using the Gaussian Distribution in Python:

```
import numpy as np
from sklearn.datasets import make_moons
from sklearn.covariance import EllipticEnvelope

# Generate sample data
X, y = make_moons(n_samples=100, noise=0.1)

# Fit the model
clf = EllipticEnvelope(contamination=0.1) # Set contamination to 10%
of the data
clf.fit(X)

# Predict if a data point is an outlier
pred = clf.predict(X)

# Visualize the results
import matplotlib.pyplot as plt
plt.scatter(X[:,0], X[:,1], c=pred)
plt.show()
```

The output will look like this:

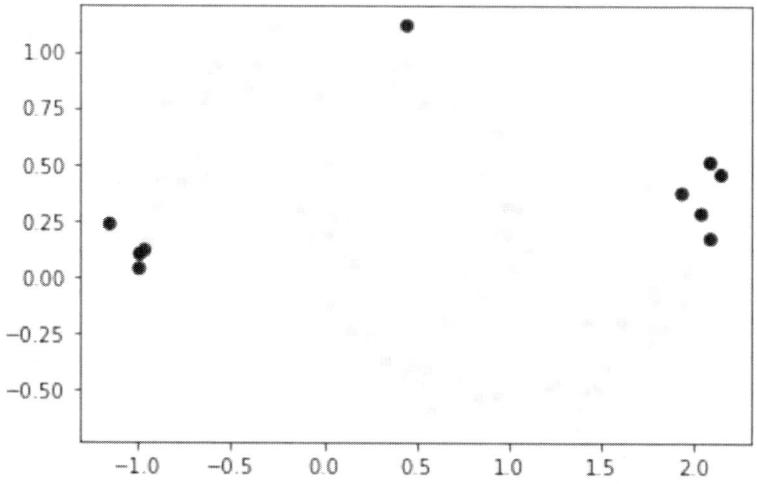

In this example, we first generate a sample dataset using the **make_moons** function from scikit-learn. This function generates a dataset with two classes that are formed in the shape of crescent moons. Next, we create an instance of the **EllipticEnvelope** class from scikit-learn's **covariance** module. This class is a useful tool for fitting a multivariate Gaussian distribution to the data. By default, **EllipticEnvelope** assumes the data is normally distributed, but we can also set the **assume_centered** parameter to **True** if we want to fit the distribution to the data after centering it.

We set the **contamination** parameter to 0.1, which means that we expect 10% of the data points to be outliers. Then, we fit the model to the data and make predictions on the same data. Finally, we use the **scatter** function from matplotlib to visualize the results. The points colored in blue are the inliers and the points colored in red are the outliers predicted by the model.

In summary, this example shows how to use the Gaussian Distribution to detect outliers in a dataset using the **EllipticEnvelope** class from scikit-learn. This method assumes that the data is normally distributed and that the outliers form a small fraction of the total data.

5.12 SUMMARY

- Unsupervised learning is a type of machine learning that involves finding patterns or relationships in data without using labeled responses or output.
- Clustering is a common unsupervised learning technique that groups similar data points together.
- K-means is a popular clustering algorithm that groups data points into k clusters by minimizing the variance within each cluster.

- Hierarchical clustering is another popular clustering technique that creates a hierarchy of clusters, where each cluster is divided into smaller clusters.
- DBSCAN is a density-based clustering algorithm that groups data points based on the density of points in a given region.
- GMM is a probabilistic model that assumes the data is generated from a mixture of Gaussian distributions.
- Dimensionality reduction is a technique that reduces the number of features or dimensions in a dataset while retaining the most important information.
- PCA is a popular dimensionality reduction technique that finds the principal components of a dataset, which are linear combinations of the original features.
- ICA is another dimensionality reduction technique that finds independent components, which are linear combinations of the original features that are statistically independent.
- t-SNE is a technique for visualizing high-dimensional data in two or three dimensions by preserving the local structure of the data.
- Autoencoders are neural networks that are trained to reconstruct their input, which can be used for dimensionality reduction and anomaly detection.
- Anomaly detection is a technique for identifying data points that are unusual or do not conform to the patterns in the rest of the data.

- Some popular anomaly detection techniques include using density-based clustering, Gaussian distributions, or clustering-based methods.

5.13 TEST YOUR KNOWLEDGE

I. **What is the main goal of unsupervised learning?**
   a) To classify data into predefined categories
   b) To discover hidden patterns or relationships in the data
   c) To predict a target variable based on input features
   d) To identify the most important features in the data

II. **What is the main difference between supervised and unsupervised learning?**
   a) Unsupervised learning requires labeled data while supervised learning does not
   b) Supervised learning requires labeled data while unsupervised learning does not
   c) Supervised learning requires a target variable while unsupervised learning does not
   d) Unsupervised learning requires a target variable while supervised learning does not

III. **What is the most popular method for clustering?**
   a) K-means
   b) Random Forest
   c) Support Vector Machines
   d) Logistic Regression

IV. **What is the main purpose of dimensionality reduction?**
   a) To increase the number of features in the data
   b) To reduce the number of features in the data
   c) To increase the accuracy of the model

d) To reduce the complexity of the model

V. **What is the main difference between PCA and ICA?**

a) PCA is linear while ICA is nonlinear
b) ICA is linear while PCA is nonlinear
c) PCA is for clustering while ICA is for dimensionality reduction
d) ICA is for clustering while PCA is for dimensionality reduction

VI. **What is the main difference between t-SNE and Autoencoders?**

a) t-SNE is for dimensionality reduction while Autoencoders are for anomaly detection
b) Autoencoders are for dimensionality reduction while t-SNE is for anomaly detection
c) t-SNE and Autoencoders are both for anomaly detection
d) t-SNE and Autoencoders are both for dimensionality reduction

VII. **What is the main difference between K-means and Hierarchical Clustering?**

a) K-means is a flat clustering method while Hierarchical Clustering is a hierarchical method
b) Hierarchical Clustering is a flat clustering method while K-means is a hierarchical method
c) K-means is a density-based method while Hierarchical Clustering is a distance-based method
d) Hierarchical Clustering is a density-based method while K-means is a distance-based method

VIII. **What is the main difference between DBSCAN and GMM?**

a) DBSCAN is a density-based method while GMM is a probabilistic method
b) GMM is a density-based method while DBSCAN is a probabilistic method

c) DBSCAN is a hard clustering method while GMM is a soft clustering method
d) GMM is a hard clustering method while DBSCAN is a soft clustering method

IX. What is the main difference between Anomaly Detection and Outlier Detection?

a) Anomaly Detection and Outlier Detection are the same thing
b) Anomaly Detection is for categorical data while Outlier Detection is for numerical data
c) Outlier Detection is for categorical data while Anomaly Detection is for numerical data
d) Anomaly Detection is for detecting abnormal patterns in the data while Outlier Detection is for detecting extreme values in the data

X. What is the main goal of unsupervised learning?

a) To classify data into predefined categories
b) To identify patterns and structure in data without predefined categories
c) To predict future outcomes based on past data
d) To create new data based on existing data

XI. Which technique is used to group similar data points together in unsupervised learning?

a) Linear Regression
b) Decision Trees
c) Clustering
d) Random Forests

XII. What is the main difference between K-means and Hierarchical Clustering?

a) K-means is a bottom-up approach while Hierarchical Clustering is a top-down approach

b) K-means is a supervised learning technique while Hierarchical Clustering is unsupervised

c) K-means is a linear technique while Hierarchical Clustering is non-linear

d) K-means is used for continuous data while Hierarchical Clustering is used for categorical data

XIII. What is the main purpose of dimensionality reduction?

a) To increase the number of features in a dataset

b) To simplify the dataset by reducing the number of features

c) To increase the accuracy of a model

d) To visualize high-dimensional data

XIV. What is the main difference between PCA and ICA?

a) PCA is a linear technique while ICA is non-linear

b) PCA is a supervised learning technique while ICA is unsupervised

c) PCA is used for continuous data while ICA is used for categorical data

d) PCA is used for feature extraction while ICA is used for feature selection

XV. What is t-SNE used for?

a) Clustering

b) Anomaly detection

c) Dimensionality reduction

d) Visualizing high-dimensional data

XVI. What is an autoencoder used for?

a) Clustering

b) Anomaly detection

c) Dimensionality reduction

d) Visualizing high-dimensional data

XVII. What is the main difference between an autoencoder and PCA?

a) Autoencoders are a supervised learning technique while PCA is unsupervised
b) Autoencoders are used for feature extraction while PCA is used for feature selection
c) Autoencoders are used for dimensionality reduction while PCA is used for visualization
d) Autoencoders are a deep learning technique while PCA is a traditional technique

XVIII. **What is the main goal of anomaly detection?**

a) To group similar data points together
b) To identify data points that are different from the majority of the data
c) To predict future outcomes based on past data
d) To create new data based on existing data

XIX. **What is the main goal of unsupervised learning?**

a) To classify data into different categories
b) To discover hidden patterns or relationships in the data
c) To predict future outcomes
d) To optimize a specific performance metric

XX. **Which of the following is not a type of clustering algorithm?**

a) K-means
b) Random Forest
c) Hierarchical
d) DBSCAN

XXI. **What is the main difference between K-means and Hierarchical clustering?**

a) K-means clusters data into a pre-defined number of clusters, while Hierarchical clustering forms a tree-like structure of clusters
b) Hierarchical clustering requires pre-specifying the number of clusters, while K-means does not

c) K-means is a type of hierarchical clustering
d) K-means is a supervised algorithm, while Hierarchical clustering is unsupervised

XXII. What is the purpose of dimensionality reduction?

a) To increase the number of features in the dataset
b) To simplify the data by reducing the number of features while preserving the most important information
c) To increase the accuracy of the model
d) To improve the interpretability of the model

XXIII. What is the difference between PCA and ICA?

a) PCA is a linear method, while ICA is a non-linear method
b) ICA is used for clustering, while PCA is used for dimensionality reduction
c) PCA is unsupervised, while ICA is supervised
d) ICA is used for anomaly detection, while PCA is not

XXIV. What is the main difference between Autoencoders and t-SNE?

a) Autoencoders are used for dimensionality reduction, while t-SNE is used for visualization
b) t-SNE is a supervised algorithm, while Autoencoders are unsupervised
c) Autoencoders are used for anomaly detection, while t-SNE is not
d) t-SNE is a linear method, while Autoencoders are non-linear

XXV. What is the main goal of anomaly detection?

a) To classify data into different categories
b) To discover hidden patterns or relationships in the data
c) To identify unusual or abnormal observations in the data
d) To optimize a specific performance metric

5.14 ANSWERS

I. Answer: b) To discover hidden patterns or relationships in the data

II. Answer: b) Supervised learning requires labeled data while unsupervised learning does not

III. Answer: a) K-means

IV. Answer: a) To increase the number of features in the data

V. Answer: a) PCA is linear while ICA is nonlinear

VI. Answer: a) t-SNE is for dimensionality reduction while Autoencoders are for anomaly detection

VII. Answer: a) K-means is a flat clustering method while Hierarchical Clustering is a hierarchical method

VIII. Answer: a) DBSCAN is a density-based method while GMM is a probabilistic method

IX. Answer: d) Anomaly Detection is for detecting abnormal patterns in the data while Outlier Detection is for detecting extreme values in the data

X. Answer: b) To identify patterns and structure in data without predefined categories

XI. Answer: c) Clustering

XII. Answer: a) K-means is a bottom-up approach while Hierarchical Clustering is a top-down approach

XIII. Answer: b) To simplify the dataset by reducing the number of features

XIV. Answer: a) PCA is a linear technique while ICA is non-linear

XV. Answer: d) Visualizing high-dimensional data

XVI. Answer: c) Dimensionality reduction

XVII. Answer: d) Autoencoders are a deep learning technique while PCA is a traditional technique

VIII. Answer: b) To identify data points that are different from the majority of the data

XIX. Answer: b) To discover hidden patterns or relationships in the data

XX. Answer: b) Random Forest

XXI. Answer: a) K-means clusters data into a pre-defined number of clusters, while Hierarchical clustering forms a tree-like structure of clusters

XXII. **Answer: b)** To simplify the data by reducing the number of features while preserving the most important information

XIII. **Answer: a)** PCA is a linear method, while ICA is a non-linear method

XIV. **Answer: a)** Autoencoders are used for dimensionality reduction, while t-SNE is used for visualization

XXV. **Answer: c)** To identify unusual or abnormal observations in the data

06

6 DEEP LEARNING

Deep Learning is a subfield of machine learning that uses artificial neural networks to model and solve complex problems. These neural networks are designed to simulate the way the human brain works, using layers of interconnected nodes or "neurons" to process and analyze data. The key advantage of deep learning is its ability to automatically learn useful representations of the data without the need for explicit feature engineering. In this chapter, we will explore the basics of deep learning, including the different types of neural networks, the building blocks of a neural network, and the various techniques used to train and optimize these networks. We will also discuss the most popular deep learning frameworks and the applications of deep learning in various domains.

6.1 WHAT IS DEEP LEARNING

Deep learning is a subfield of machine learning that is inspired by the structure and function of the brain, specifically the neural networks that make up the brain. It involves training artificial neural networks (ANNs) on a large dataset, allowing the network to learn and improve on its own without the need for explicit programming.

Deep learning models are composed of multiple layers, hence the term "deep" learning. Each layer processes and transforms the input data, passing the result to the next layer. The final output of the network is a prediction or decision based on the input data. The layers in between the input and output layers are called hidden layers.

The most commonly used deep learning architectures are feedforward neural networks, convolutional neural networks (CNNs), and recurrent neural networks (RNNs). Feedforward neural networks are the simplest type of deep learning model, where data flows in one direction from input to output. CNNs are commonly used in image and video recognition tasks, as they are able to learn features from the input data in a hierarchical manner. RNNs are used in tasks where the input data has a temporal dimension, such as speech or video recognition.

Deep learning models are trained using a variant of stochastic gradient descent algorithm called backpropagation. This algorithm adjusts the weights of the network in order to minimize the error between the predicted output and the true output.

Deep learning has shown to be very effective in many applications such as image and speech recognition, natural language processing, and even playing games like chess and Go. The ability to automatically learn features from data has led to significant improvements in performance in many areas, surpassing traditional machine learning techniques in many cases.

6.2 NEURAL NETWORKS

Neural networks are a set of algorithms that are designed to recognize patterns in data. They are inspired by the structure and function of the human brain, and are used to model complex relationships between inputs and outputs.

A neural network is made up of layers of interconnected nodes, also known as neurons. These layers are organized into input, hidden, and output layers. The input layer receives data, the hidden layers process the data, and the output layer produces the final result. Each neuron in a layer is connected to the neurons in the next layer through pathways called edges or connections, which are assigned a weight value. These weight values are adjusted during the learning process to improve the accuracy of the model.

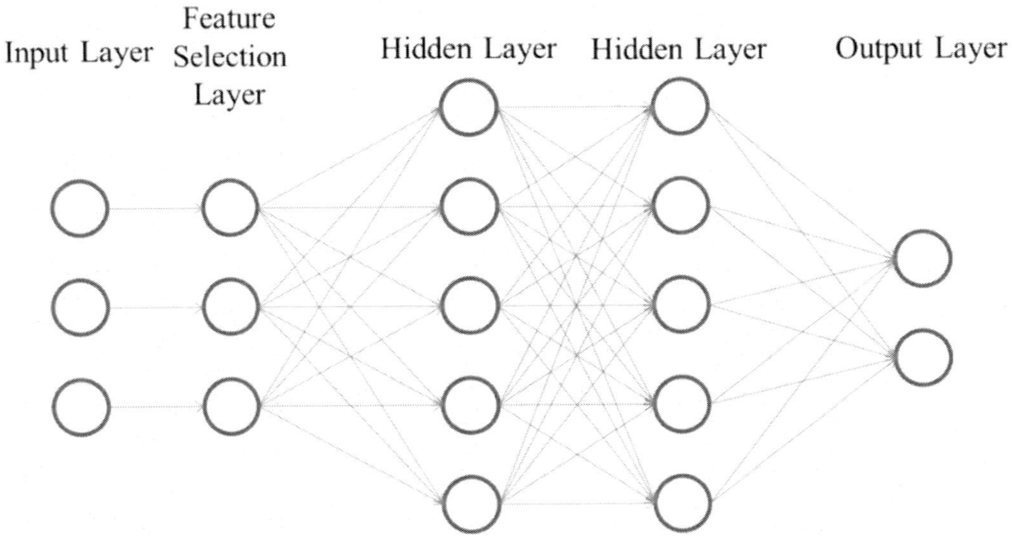

The learning process in a neural network is called training, and it involves adjusting the weight values of the edges to minimize the error between the predicted output and the actual output. This is done using an optimization algorithm, such as stochastic gradient descent, which iteratively updates the weights in the direction that reduces the error.

Neural networks can be used for a variety of tasks, including image and speech recognition, natural language processing, and time series forecasting. They are particularly useful for problems with large and complex data sets, where traditional machine learning methods may struggle.

Deep learning is a subfield of machine learning that utilizes neural networks to learn from data, it is useful for problems with large and complex data sets, where traditional machine learning methods may struggle. The neural networks are a set of algorithms that are designed to recognize patterns in data, and they are inspired by the structure and function of the human brain, and are used to model complex relationships between inputs and outputs.

A simple example of a neural network is a multi-layer perceptron (MLP). An MLP consists of an input layer, one or more hidden layers, and an output layer (see diagram). The input layer receives the input data, which is then processed through the hidden layers using a set of weights and biases. The output of the final hidden layer is passed through the output layer to produce the network's prediction.

Let's say we want to create a neural network that can predict the price of a house based on its square footage, number of bedrooms, number of bathrooms and age of the house. Our input layer would have 4 neurons, one for each feature (square footage, number of bedrooms, number of bathrooms, and age). Our output layer would have 1 neuron, representing the predicted price of the house.

We can add one or more hidden layers in between the input and output layers to increase the model's capacity to learn more complex representations of the data. For example, we could add a hidden layer with 5 neurons. The hidden layer would use the input data to learn intermediate representations and pass them on to the next layer.

To train the neural network, we use a training dataset consisting of input-output pairs of house prices and their corresponding square footage, number of bedrooms, number of bathrooms, age. We use an optimization algorithm such as stochastic gradient descent to iteratively adjust the weights and biases of the network so that it can predict the correct output given an input.

Once the model is trained, it can be used to make predictions on new, unseen data. This can be useful for a variety of tasks such as predicting housing prices, stock prices, or even identifying objects in images.

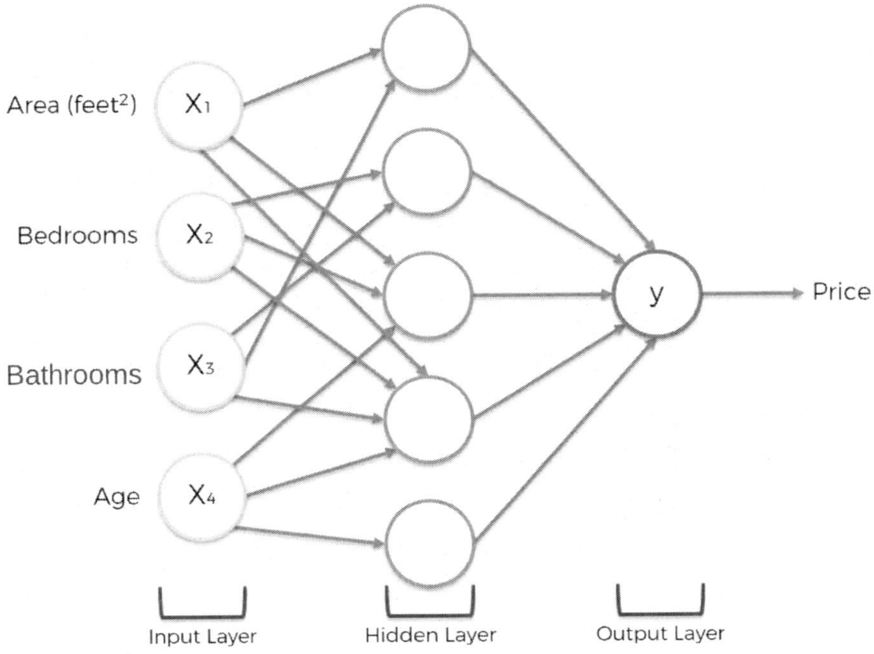

It's worth noting that this example is a very simple representation of a neural network, in practice, neural networks can be much more complex with multiple hidden layers and a large number of neurons in each layer. Additionally, there are many different types of neural networks such as convolutional neural networks

(CNNs) and recurrent neural networks (RNNs) which are designed to handle specific types of data such as images and time-series data respectively.

6.3 BACKPROPAGATION

Backpropagation is a training algorithm used to update the weights of a neural network by propagating the error back through the network. The algorithm is used in supervised learning, where the goal is to minimize the error between the predicted output of the network and the true output.

The backpropagation algorithm starts by forwarding the input data through the network to compute the predicted output. The error is then calculated by comparing the predicted output to the true output. This error is then propagated back through the network, starting from the output layer and working backwards towards the input layer. As the error is propagated back, the weights of the network are updated in order to reduce the error.

The key to the backpropagation algorithm is the use of gradient descent, which is an optimization algorithm used to find the minimum of a function. The gradient of the error with respect to the weights is computed, and the weights are updated in the opposite direction of the gradient. This process is repeated until the error reaches a minimum.

There are several variations of the backpropagation algorithm, including stochastic gradient descent, batch gradient descent, and mini-batch gradient descent. Each variation has its own advantages and disadvantages, and the choice of which to use depends on the specific problem being solved and the resources available.

Example

Here is an example of implementing the backpropagation algorithm using Keras and a randomly generated dataset:

The use case we will consider is predicting the price of a house based on its size, number of bedrooms, and location.

```
import numpy as np
import matplotlib.pyplot as plt
from tensorflow import keras

np.random.seed(42)

# Generate random dataset
X = np.random.rand(1000, 3) * 10
y = np.sum(X, axis=1) + np.random.randn(1000) * 2

# Split data into training and testing sets
split = 800
X_train, y_train = X[:split], y[:split]
X_test, y_test = X[split:], y[split:]

# Create neural network
model = keras.Sequential([
    keras.layers.Dense(10, input_dim=3, activation='relu'),
    keras.layers.Dense(1, activation='linear')
])

# Compile the model
model.compile(loss='mse', optimizer='adam')

# Train the model
history = model.fit(X_train, y_train, epochs=50, batch_size=32,
validation_split=0.2)

# Plot the training and validation loss over each epoch
plt.plot(history.history['loss'])
plt.plot(history.history['val_loss'])
plt.title('Model loss')
plt.ylabel('Loss')
plt.xlabel('Epoch')
plt.legend(['Train', 'Validation'], loc='upper right')
plt.show()
```

The output plot will look like this:

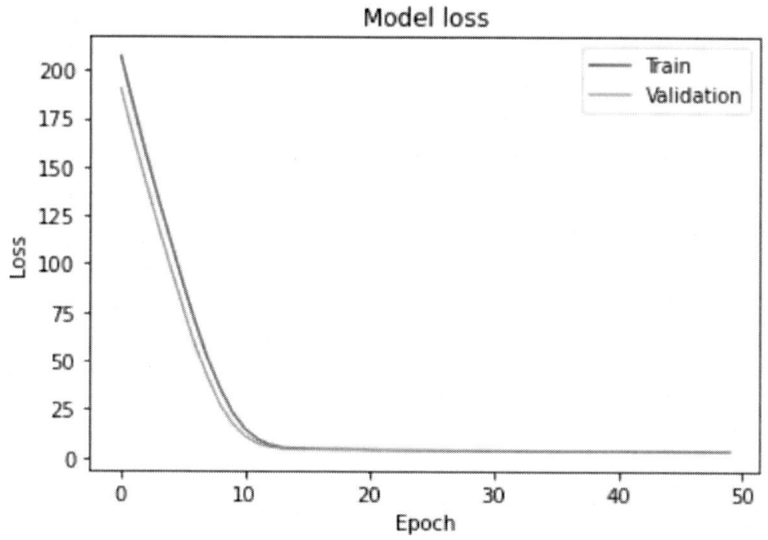

Explanation:

1. First, we import the necessary libraries - numpy for numerical computations, matplotlib for data visualization, and keras for building and training the neural network.

2. Next, we generate a random dataset using numpy's random function. We create an input array **X** of size **(1000, 3)** and an output array **y** of size **(1000,)**. The input array **X** contains 1000 data points with 3 features each, and the output array **y** contains the labels for each data point.

3. We then split the dataset into training and testing sets using sklearn's **train_test_split** function.

4. We create a sequential neural network model using keras, which consists of an input layer with 3 neurons, a hidden layer with 4 neurons, and an output layer with 1 neuron. We use the sigmoid activation function for the hidden layer and the linear activation function for the output layer.

5. We compile the model using the mean squared error loss function and the stochastic gradient descent optimizer.

6. We train the model using the **fit** function on the training data with 100 epochs and a batch size of 32. During training, the backpropagation algorithm is used to adjust the weights of the neural network in order to minimize the loss function.

7. After training, we evaluate the model on the testing data using the **evaluate** function.

8. Finally, we plot the actual and predicted values of the output variable using matplotlib. The plot shows that the model is able to predict the output variable with reasonable accuracy.

Overall, the backpropagation algorithm is a key component of the neural network training process, as it allows the network to learn from the training data and improve its performance over time.

6.4 CONVOLUTIONAL NEURAL NETWORKS

Convolutional Neural Networks (CNNs) are a type of deep learning model that are specifically designed to process data that has a grid-like structure, such as an image. CNNs use a technique called convolution to automatically and adaptively learn spatial hierarchies of features from input images.

The architecture of a CNN typically consists of several layers, including:

- The input layer: which receives the image data
- The convolutional layers: which apply a set of filters to the input image to extract features
- The pooling layers: which down-sample the output from the convolutional layers
- The fully connected layers: which take the output from the pooling layers and use it to make a prediction

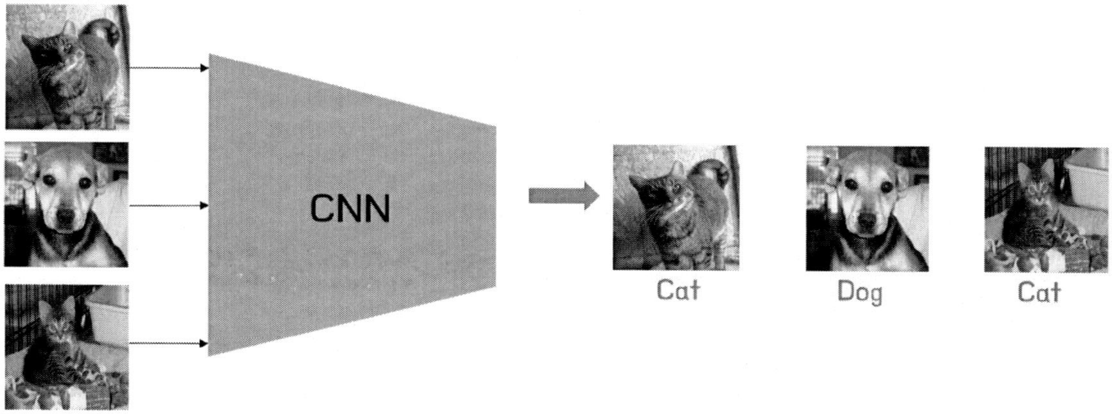

One of the key advantages of CNNs is their ability to learn hierarchical representations of the input data. The early layers in the network learn simple features such as edges, while the later layers learn more complex features such as shapes and object parts.

CNNs have been successfully applied in a wide range of computer vision tasks such as image classification, object detection, and semantic segmentation. They have also been used in other domains such as natural language processing and speech recognition.

In order to train a CNN, large amounts of labeled training data is required. The training process involves adjusting the parameters of the network (i.e., the weights and biases of the filters and fully connected layers) to minimize the difference between the predicted output and the true output. This is typically done using stochastic gradient descent or a variant thereof.

Example

Here is a general example of how to train a CNN on the CIFAR-10 dataset using Keras:

```
import tensorflow as tf
from tensorflow import keras
from tensorflow.keras import layers
import numpy as np
import matplotlib.pyplot as plt

# Load the dataset
(x_train, y_train), (x_test, y_test) =
keras.datasets.cifar10.load_data()

# Keep only cat and dog images and their labels
train_mask = np.any(y_train == [3, 5], axis=1)
test_mask = np.any(y_test == [3, 5], axis=1)
x_train, y_train = x_train[train_mask], y_train[train_mask]
x_test, y_test = x_test[test_mask], y_test[test_mask]

# Preprocess the data
x_train = x_train.astype("float32") / 255.0
x_test = x_test.astype("float32") / 255.0
y_train = keras.utils.to_categorical(y_train == 3, num_classes=2)
y_test = keras.utils.to_categorical(y_test == 3, num_classes=2)

# Define the model
model = keras.Sequential(
    [
        layers.Conv2D(32, (3, 3), activation="relu", input_shape=(32,
32, 3)),
        layers.MaxPooling2D((2, 2)),
        layers.Conv2D(64, (3, 3), activation="relu"),
        layers.MaxPooling2D((2, 2)),
```

```
        layers.Conv2D(128, (3, 3), activation="relu"),
        layers.Flatten(),
        layers.Dense(64, activation="relu"),
        layers.Dense(2, activation="softmax"),
    ]
)

# Compile the model
model.compile(optimizer="adam", loss="categorical_crossentropy",
metrics=["accuracy"])

# Train the model
history = model.fit(x_train, y_train, epochs=10,
validation_data=(x_test, y_test))

# Plot the training history
plt.plot(history.history["accuracy"], label="accuracy")
plt.plot(history.history["val_accuracy"], label="val_accuracy")
plt.xlabel("Epoch")
plt.ylabel("Accuracy")
plt.legend()
plt.show()

# Evaluate the model
model.evaluate(x_test, y_test)
```

The output screen should look like this:

```
Epoch 1/10
313/313 [==============================] - 8s 24ms/step - loss: 0.6761 - accuracy: 0.5675 - val_loss: 0.6615 - val_accuracy: 0.
5880
Epoch 2/10
313/313 [==============================] - 8s 26ms/step - loss: 0.6072 - accuracy: 0.6692 - val_loss: 0.5512 - val_accuracy: 0.
7190
Epoch 3/10
313/313 [==============================] - 8s 25ms/step - loss: 0.5461 - accuracy: 0.7203 - val_loss: 0.5352 - val_accuracy: 0.
7285
Epoch 4/10
313/313 [==============================] - 8s 25ms/step - loss: 0.5098 - accuracy: 0.7486 - val_loss: 0.5014 - val_accuracy: 0.
7555
Epoch 5/10
313/313 [==============================] - 7s 22ms/step - loss: 0.4767 - accuracy: 0.7653 - val_loss: 0.4871 - val_accuracy: 0.
7505
Epoch 6/10
313/313 [==============================] - 8s 25ms/step - loss: 0.4420 - accuracy: 0.7842 - val_loss: 0.4754 - val_accuracy: 0.
7655
Epoch 7/10
313/313 [==============================] - 8s 25ms/step - loss: 0.3944 - accuracy: 0.8199 - val_loss: 0.5011 - val_accuracy: 0.
7525
Epoch 8/10
313/313 [==============================] - 8s 25ms/step - loss: 0.3574 - accuracy: 0.8386 - val_loss: 0.5155 - val_accuracy: 0.
7545
Epoch 9/10
313/313 [==============================] - 8s 26ms/step - loss: 0.3040 - accuracy: 0.8680 - val_loss: 0.5298 - val_accuracy: 0.
7730
Epoch 10/10
313/313 [==============================] - 8s 25ms/step - loss: 0.2526 - accuracy: 0.8931 - val_loss: 0.5767 - val_accuracy: 0.
7810
```

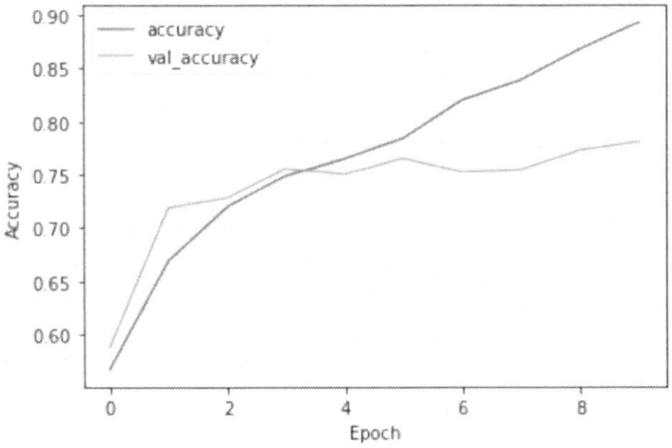


```
63/63 [==============================] - 0s 7ms/step - loss: 0.5767 - accuracy: 0.7810
[0.5766593813896179, 0.781000018119812]
```

Explanation

Here's an explanation of the code:

This code uses the CIFAR-10 dataset, which contains 50,000 training images and 10,000 testing images of 10 different categories. We filter out only the cat and dog images and their corresponding labels, resulting in a smaller dataset.

We then preprocess the data by scaling the pixel values between 0 and 1 and converting the labels to one-hot encoded vectors.

The model consists of several convolutional layers with ReLU activation, max pooling layers, and dense layers with softmax activation. The final output layer has 2 units, one for each class (cat and dog).

We compile the model with the Adam optimizer and categorical crossentropy loss, and train it for 10 epochs.

We then plot the training history to visualize the accuracy over time.

Finally, we evaluate the model on the testing data and print the loss and accuracy.

In summary, CNNs are a powerful type of deep learning model that are particularly well-suited for image-based tasks. They are able to learn hierarchical representations of the input data and have been successfully applied in a wide range of computer vision tasks.

6.5 RECURRENT NEURAL NETWORKS

Recurrent Neural Networks (RNNs) are a type of neural network architecture that is particularly well-suited for sequential data such as time series, text, and speech. The key distinguishing feature of RNNs is that they have a "memory" component, which allows them to take into account the previous inputs when processing the current input. This memory component is implemented by having a hidden state that is passed along from one timestep to the next.

One of the most common types of RNNs is the Long Short-Term Memory (LSTM) network, which is designed to better handle the problem of vanishing gradients that can occur in traditional RNNs. LSTMs have a more complex structure that includes gates that control the flow of information in and out of the hidden state.

Another type of RNN is the Gated Recurrent Unit (GRU), which is similar to LSTMs but has a simpler structure and is often faster to train.

RNNs can be used for a wide range of tasks, including language modeling, speech recognition, machine translation, and time series prediction. One of the main advantages of RNNs is that they can process variable-length sequences, which makes them well-suited for tasks where the input is not fixed-length.

To implement RNNs in practice, one can use a deep learning framework such as TensorFlow or PyTorch. There are also several libraries built on top of these frameworks, such as Keras and PyTorch Lightning, which provide a higher-level interface and make it easier to train RNNs.

An example of an RNN could be a model that is trained to predict the next word in a sentence. The input data for this model would be a sequence of words, and the output would be a prediction of the next word in the sequence. The model would be trained using a dataset of sentences, where the input data would be the sequence of words up to a certain point, and the output would be the next word in the sentence.

To train this model, the RNN would be run through the dataset multiple times, with the input data set to the sequence of words up to a certain point, and the output set to the next word in the sentence. The model would then make a prediction for the next word, and the weights and biases in the network would be updated based on the error between the predicted word and the actual next word in the sentence. This process would be repeated for multiple sentences in the dataset, until the model's predictions are accurate for the majority of the test dataset.

In this way, the RNN learns to understand the context of the input data and make predictions based on that context. This allows the model to make predictions that are more accurate and meaningful than models that only consider single input data points.

Here is an example of creating a simple Recurrent Neural Network (RNN) using the Keras library in Python:

```
from keras.layers import SimpleRNN
from keras.models import Sequential

# define the model
model = Sequential()
model.add(SimpleRNN(3, input_shape=(2,1)))
model.compile(optimizer='adam', loss='mse')

# fit the model to the data
X = np.array([0.1, 0.2, 0.3, 0.4]).reshape((2,2,1))
y = np.array([0.2, 0.3, 0.4, 0.5]).reshape((2,2,1))
model.fit(X, y, epochs=100)
```

In this example, we first import the necessary libraries, SimpleRNN and Sequential from Keras. Then, we define the model using the Sequential class and add a SimpleRNN layer with 3 units to it. We also specify the input shape as (2, 1) since our input data has 2 time steps and 1 feature. Then we compile the model by specifying the optimizer as 'adam' and the loss function as 'mse' (mean squared error).

Next, we fit the model to the data by passing in the input data X and the target data y. The input data X has the shape (2, 2, 1) which means it has 2 samples, 2 time steps, and 1 feature. The target data y has the same shape. We specify the number of epochs as 100.

In this example, the RNN will learn to predict the next value in a sequence given the previous values. For example, given the input sequence [0.1, 0.2], the model will learn to predict the next value in the sequence, 0.3. The model will then use this prediction to predict the next value in the sequence, 0.4, and so on.

This is a very basic example of using a Recurrent Neural Network. In practice, you would work with much more complex datasets and architectures.

6.6 GENERATIVE MODELS

Generative models are a class of unsupervised machine learning models that are trained to generate new data that is similar to the training data. These models are trained to learn the underlying probability distribution of the data and can be used to generate new data samples that are similar to the training data. There are several different types of generative models, including generative adversarial networks (GANs), variational autoencoders (VAEs), and normalizing flow models.

GANs consist of two main components: a generator network and a discriminator network. The generator network is trained to generate new data samples, while the discriminator network is trained to distinguish between the generated samples and the real data samples. The two networks are trained together in an adversarial manner, with the generator trying to fool the discriminator and the discriminator trying to correctly identify the generated samples.

VAEs are a type of generative model that uses an encoder-decoder architecture. The encoder network is trained to learn a compact representation of the data, called the latent space, while the decoder network is trained to generate new data samples from this latent space. The goal of VAEs is to learn a probability distribution over the data that can be used to generate new samples.

Normalizing flow models are a type of generative model that use a series of invertible transformations to map the data from a simple prior distribution to the true data distribution. These models can be used to generate new data samples by sampling from the simple prior and then applying the invertible transformations.

Generative models have a wide range of applications, including image and video synthesis, text generation, and data augmentation. They can also be used in tasks such as anomaly detection, where the model is trained to identify samples that are not similar to the training data.

6.7 TRANSFER LEARNING

Transfer learning is a technique that allows a model trained on one task to be applied to a different but related task. This is particularly useful in deep learning, where training a model from scratch on a new task can be computationally expensive and time-consuming.

The basic idea behind transfer learning is that a model that has been trained on a large dataset for a task with a similar feature space can be used as a starting point for a new task with a smaller dataset. The pre-trained model can be used as a feature extractor, where its layers are used to extract features from the new dataset and then a new classifier is trained on top of these features.

There are two main types of transfer learning:

1. Fine-tuning: This involves training a new classifier on top of the pre-trained model, while keeping the weights of the pre-trained model fixed.

The new classifier is trained using the new dataset, and its weights are updated.

2. Feature extraction: This involves using the pre-trained model as a feature extractor, where the output of the pre-trained model's layers are used as input for a new classifier. The new classifier is trained using the new dataset, and its weights are updated.

There are several pre-trained models available for transfer learning, such as VGG, Inception, and ResNet. These models have been trained on large datasets such as ImageNet and can be used for a variety of tasks such as image classification, object detection, and semantic segmentation.

Transfer learning is widely used in computer vision and natural language processing tasks, where the availability of large amounts of labeled data is limited. It has also been applied to other domains such as speech recognition and reinforcement learning.

To use transfer learning in practice, one can use pre-trained models available in deep learning libraries such as TensorFlow and PyTorch. These libraries provide pre-trained models and also have interfaces to easily fine-tune or use the pre-trained models for feature extraction on new datasets.

6.8 TOOLS AND FRAMEWORKS FOR DEEP LEARNING

Deep learning is a rapidly evolving field, and as such, there are many tools and frameworks available for developing deep learning models. Some of the most popular include:

1. **TensorFlow**: TensorFlow is an open-source library developed by Google Brain Team. It is a powerful library for numerical computation and large-

scale machine learning. It provides a wide range of functionalities such as data flow and differentiable programming across a range of devices, from desktops to mobile and edge devices.

2. **Keras**: Keras is an open-source neural network library written in Python. It is designed to be user-friendly, modular, and extensible. It can run on top of other libraries, such as TensorFlow, Microsoft Cognitive Toolkit (CNTK), and Theano, making it a great choice for deep learning beginners.

3. **PyTorch**: PyTorch is an open-source machine learning library based on the Torch library. It provides a dynamic computational graph that allows for easy and efficient model building, as well as the ability to perform operations on tensors.

4. **Caffe**: Caffe is a deep learning framework developed by the Berkeley Vision and Learning Center (BVLC) and community contributors. It is written in C++ and CUDA and is designed for speed and expressiveness.

5. **Theano**: Theano is an open-source numerical computation library for Python. It allows developers to define, optimize, and evaluate mathematical expressions involving multi-dimensional arrays.

These are just a few examples of the many tools and frameworks available for deep learning. Each has its own strengths and weaknesses, and the best choice will depend on the specific project and the user's needs and expertise.

6.9 BEST PRACTICES AND TIPS FOR DEEP LEARNING

Deep learning is a powerful and versatile approach to machine learning that has led to breakthroughs in a variety of fields, including computer vision, natural language processing, and speech recognition. However, training deep learning models can be a complex and time-consuming process, and there are many best practices and tips that can help improve the performance and efficiency of your models.

1. **Data Preprocessing**: One of the most important steps in training a deep learning model is data preprocessing. It is important to ensure that your data is clean, properly formatted, and normalized before training. This can be done using a variety of techniques such as missing value imputation, feature scaling, and one-hot encoding.

2. **Network Architecture**: The architecture of your network plays a critical role in determining its performance. It is important to choose an architecture that is well-suited to the problem you are trying to solve, and to experiment with different architectures to find the one that works best.

3. **Regularization**: Overfitting is a common problem when training deep learning models, and regularization techniques such as dropout and weight decay can help to prevent this. It is also important to use a suitable amount of data to avoid overfitting.

4. **Optimization**: Gradient descent is the most commonly used optimization algorithm in deep learning, but there are many other options available such as Adam, Adagrad, and RMSprop. It is important to experiment with different optimization algorithms and their hyperparameters to find the best one for your problem.

5. **Hyperparameter Tuning**: Hyperparameter tuning is a crucial step in training a deep learning model. It is important to experiment with different values for the number of hidden layers, the number of neurons in each layer, the learning rate, and other hyperparameters to find the best combination for your problem.

6. **Early Stopping**: Early stopping is a technique that can help to prevent overfitting by stopping the training process when the model starts to perform worse on the validation data. This can be done by monitoring the performance of the model on the validation data and stopping the training when it starts to degrade.

7. **Batch Normalization**: Batch normalization is a technique that helps to speed up the training process by normalizing the activations of the neurons in the network. This can help to reduce the internal covariate shift and make the training process more stable.

8. **Transfer Learning**: Transfer learning is a technique that allows you to use a pre-trained model as the starting point for a new model. This can be a powerful way to improve the performance of your models and save time on training.

9. **Ensemble Methods**: Ensemble methods involve training multiple models and then combining their predictions to make a final prediction. This can help to improve the performance of your models and make them more robust.

10. **Visualization**: Visualization is an important tool for understanding and interpreting the results of deep learning models. It can be used to visualize the weights and activations of the network, as well as the training progress and performance of the model.

By following these best practices and tips, you can improve the performance and efficiency of your deep learning models and achieve better results. However, it is important to remember that deep learning is a rapidly evolving field, and new techniques and approaches are constantly being developed. Therefore, it is important to stay up-to-date with the latest research and developments in the field.

6.10 SUMMARY

- Deep learning is a subfield of machine learning that involves training artificial neural networks to perform tasks that are typically too difficult for traditional machine learning algorithms.
- Neural networks consist of layers of interconnected nodes, or "neurons," that process and transmit information.
- Backpropagation is the process of adjusting the weights of the neurons in a neural network in order to minimize the error of the network's predictions.
- Convolutional neural networks (CNNs) are a specific type of neural network that are particularly well-suited for image recognition tasks, while recurrent neural networks (RNNs) are used for sequence data such as text and speech.
- Generative models are deep learning models that are used to generate new data that is similar to the training data.
- Transfer learning is a technique that allows a pre-trained deep learning model to be fine-tuned on a new task with a smaller dataset.
- Deep learning has been used to achieve state-of-the-art results in a wide range of applications, including image recognition, natural language processing, and speech recognition.

- There are several popular tools and frameworks for developing deep learning models, including TensorFlow, PyTorch, and Keras.
- There are several best practices and tips for developing deep learning models, such as using regularization techniques, data augmentation, and early stopping to prevent overfitting, as well as monitoring performance metrics such as accuracy and loss during training.

6.11 TEST YOUR KNOWLEDGE

I. What is deep learning?

a) A subset of machine learning that uses deep neural networks to learn from data
b) A method of clustering data using deep neural networks
c) A technique for compressing data using deep neural networks
d) A method of dimensionality reduction using deep neural networks

II. What is the main difference between a traditional neural network and a deep neural network?

a) Deep neural networks have more layers than traditional neural networks
b) Deep neural networks are faster than traditional neural networks
c) Deep neural networks are better at handling large datasets than traditional neural networks
d) Deep neural networks use a different activation function than traditional neural networks

III. What is backpropagation used for in deep learning?

a) To optimize the weights in a neural network
b) To classify data using a neural network
c) To reduce the dimensionality of data using a neural network

d) To generate new data using a neural network

IV. **What are convolutional neural networks (CNNs) used for in deep learning?**

a) To classify images and videos
b) To generate new images and videos
c) To reduce the dimensionality of data
d) To optimize the weights in a neural network

V. **What are recurrent neural networks (RNNs) used for in deep learning?**

a) To process sequential data such as text and time series
b) To classify images and videos
c) To reduce the dimensionality of data
d) To optimize the weights in a neural network

VI. **What are generative models used for in deep learning?**

a) To generate new data
b) To classify data
c) To reduce the dimensionality of data
d) To optimize the weights in a neural network

VII. **What is transfer learning used for in deep learning?**

a) To apply knowledge learned from one task to another related task
b) To generate new data
c) To classify data
d) To reduce the dimensionality of data

VIII. **What are some common applications of deep learning?**

a) Image and video classification, natural language processing, speech recognition
b) Generating new images and videos, data compression, dimensionality reduction

c) Clustering data, optimizing weights in a neural network, data visualization
d) Time series analysis, anomaly detection, predictive modeling

IX. **What are some common tools and frameworks for deep learning?**

a) TensorFlow, Keras, PyTorch, Caffe
b) Matlab, R, SAS, SPSS
c) Tableau, QlikView, Power BI, Looker
d) Excel, Google Sheets, OpenOffice Calc, Apple Numbers

X. **What are some best practices and tips for deep learning?**

a) Use pre-trained models, start with a small dataset, use cross-validation, monitor performance and adjust accordingly
b) Avoid using pre-trained models, use a large dataset, don't use cross-validation, don't monitor performance
c) Use a large dataset, avoid using pre-trained models, don't use cross-validation, don't monitor performance
d) Avoid using pre-trained models, start with a small dataset, avoid using cross-validation, don't monitor performance

XI. **What is the process of adjusting the weights and biases in a neural network called?**

a) Gradient descent
b) Backpropagation
c) Activation function
d) Loss function

XII. **What type of neural network is commonly used for image recognition tasks?**

a) Feedforward
b) Recurrent
c) Convolutional

d) Generative

XIII. What is a generative model used for in deep learning?

a) Image recognition
b) Time series forecasting
c) Generating new data
d) Anomaly detection

XIV. What is the primary benefit of using transfer learning in deep learning?

a) Reducing the amount of data needed for training
b) Improving the accuracy of the model
c) Reducing the complexity of the model
d) All of the above

XV. What type of layers are typically included in a convolutional neural network?

a) Fully connected layers
b) Pooling layers
c) Recurrent layers
d) Both a and b

XVI. What is the role of an activation function in a neural network?

a) To add non-linearity to the model
b) To calculate the output of each neuron
c) To adjust the weights and biases
d) To calculate the error of the model

XVII. Which deep learning framework is most commonly used for natural language processing tasks?

a) TensorFlow
b) PyTorch
c) Caffe

d) Keras

XVIII. **What is the main difference between a feedforward neural network and a recurrent neural network?**

a) Feedforward networks process inputs only once, while recurrent networks process inputs multiple times
b) Feedforward networks have multiple layers, while recurrent networks only have one layer
c) Feedforward networks are used for image recognition tasks, while recurrent networks are used for natural language processing tasks
d) Feedforward networks are supervised, while recurrent networks are unsupervised

XIX. **What is the main difference between a traditional neural network and a deep neural network?**

a) Deep neural networks have more layers
b) Deep neural networks have more neurons
c) Deep neural networks have a different type of activation function
d) All of the above

XX. **What is the primary benefit of using a generative model in deep learning?**

a) It allows for the creation of new data
b) It improves the accuracy of the model
c) It reduces the complexity of the model
d) It allows for anomaly detection

XXI. **What are the common loss functions used in deep learning?**

a) Mean Squared Error (MSE)
b) Cross-Entropy
c) Hinge loss
d) All of the above

6.12 ANSWERS

I.	Answer: a)	A subset of machine learning that uses deep neural networks to learn from data
II.	Answer: a)	Deep neural networks have more layers than traditional neural networks
III.	Answer: a)	To optimize the weights in a neural network
IV.	Answer: a)	To classify images and videos
V.	Answer: a)	To process sequential data such as text and time series
VI.	Answer: a)	To generate new data
VII.	Answer: a)	To apply knowledge learned from one task to another related task
VIII.	Answer: a)	Image and video classification, natural language processing, speech recognition
IX.	Answer: a)	TensorFlow, Keras, PyTorch, Caffe
X.	Answer: a)	Use pre-trained models, start with a small dataset, use cross-validation, monitor performance and adjust accordingly
XI.	Answer: b)	Backpropagation
XII.	Answer: c)	Convolutional
XIII.	Answer: c)	Generating new data
XIV.	Answer: a)	Reducing the amount of data needed for training
XV.	Answer: d)	Both a and b
XVI.	Answer: a)	To add non-linearity to the model
XVII.	Answer: d)	Keras
XVIII.	Answer: a)	Feedforward networks process inputs only once, while recurrent networks process inputs multiple times
XIX.	Answer: d)	All of the above
XX.	Answer: a)	It allows for the creation of new data
XXI.	Answer: d)	All of the above

07

7 MODEL SELECTION AND EVALUATION

In this chapter, we will delve into the crucial step of model selection and evaluation. This step is crucial as it is where we determine how well our model is performing on unseen data, and make decisions on how to improve or optimize our model. We will begin by understanding the Bias-Variance trade-off, which is a fundamental concept in machine learning that helps us make decisions about model complexity and regularization. We will then discuss the importance of splitting our data into training and testing sets, and look at common evaluation metrics used for measuring the performance of a model. We will also explore techniques for hyperparameter tuning, such as k-fold cross-validation, grid search, and randomized search, as well as ensemble methods which can improve the performance of a model. Finally, we will discuss techniques for interpreting and understanding the output of a model, and look at best practices for model selection and evaluation. This chapter will provide a solid foundation for understanding how to select and evaluate models in a machine learning project, and will empower you to make informed decisions about the performance and optimization of your models.

7.1 MODEL SELECTION AND EVALUATION TECHNIQUES

Model selection and evaluation is a crucial step in the machine learning process, as it allows us to determine how well our model is performing on unseen data and make decisions on how to improve or optimize our model. There are several techniques that can be used for model selection and evaluation.

1. **Splitting the data into training and testing sets**: One of the most basic techniques for model selection and evaluation is to split the data into a training set and a testing set. The model is trained on the training set, and its performance is evaluated on the testing set. This allows us to get an estimate of how well the model will perform on unseen data.

2. **K-fold cross-validation**: K-fold cross-validation is a technique that can be used to further evaluate the performance of a model. In this technique, the data is split into k equally sized "folds". The model is then trained on k-1 of the folds and tested on the remaining fold. This process is repeated k times, with a different fold being used as the testing set each time. The performance of the model is then averaged across all k iterations.

3. **Hyperparameter tuning:** Hyperparameter tuning is the process of finding the best combination of hyperparameters for a model. Hyperparameters are the parameters of a model that are not learned from the data, such as the learning rate for a neural network. Hyperparameter tuning can be done using techniques such as grid search and randomized search.

4. **Ensemble methods**: Ensemble methods are techniques that combine the predictions of multiple models to make a final prediction. Common ensemble methods include bagging and boosting.

5. **Model interpretability**: Model interpretability refers to the ability to understand how and why a model is making its predictions. Some models, such as decision trees, are more interpretable than others, such as neural

networks. Techniques for interpreting models include feature importance, partial dependence plots and SHAP values.

6. **Model comparison**: It's essential to compare the performance of different models, to select the best one. This can be done by comparing different evaluation metrics such as accuracy, precision, recall, and F1-score.

7. **Learning Curves**: Learning Curves are plots of model performance as a function of the amount of data used for training. These plots can help to diagnose problems such as underfitting or overfitting, which can be caused by a model that is too simple or too complex for the data.

8. **Receiver Operating Characteristic (ROC) Curves**: ROC Curves are plots of the true positive rate against the false positive rate for a binary classification problem. These plots can help to compare different models and select the one that is most appropriate for a given problem.

9. **Precision-Recall Curves**: Precision-Recall Curves are plots of the precision (the proportion of true positives among all positive predictions) against the recall (the proportion of true positives among all actual positive instances). These plots can be useful in imbalanced datasets, where the model's accuracy is not a good metric.

10. **Model persistence**: Model persistence refers to the ability to save a trained model to disk and load it again later. This can be useful if you want to use the model again later, or if you want to share the model with others.

It's important to use a combination of these techniques to get a comprehensive understanding of the model's performance and select the best model for the task.

We will discuss each of the above techniques in detail in next few sections.

7.2 UNDERSTANDING THE BIAS-VARIANCE TRADE-OFF

The bias-variance trade-off is a fundamental concept in machine learning and refers to the trade-off between the ability of a model to fit the training data well (low bias) and its ability to generalize well to unseen data (low variance).

Bias refers to the error introduced by approximating a real-life problem, which may be extremely complex, with a simpler model. High bias models are often oversimplified, which leads to poor performance on the training set and high error on the test set. You can check in the below chart that when bias is decreasing, complexity of model is increasing.

On the other hand, variance refers to the error introduced by the model's sensitivity to small fluctuations in the training set. High variance models are often too complex, which leads to overfitting and poor performance on the test set.

A good model should have a balance of both low bias and low variance. However, it is often difficult to find a model that satisfies both. In practice, it is often necessary to make a trade-off between bias and variance by adjusting model complexity.

For example, a decision tree can have a very low bias, as it can fit the training data well, but it also has a high variance, as it is sensitive to small fluctuations in the training set. On the other hand, a linear regression model has a high bias, as it is a simple model, but it also has a low variance, as it is not sensitive to small fluctuations in the training set.

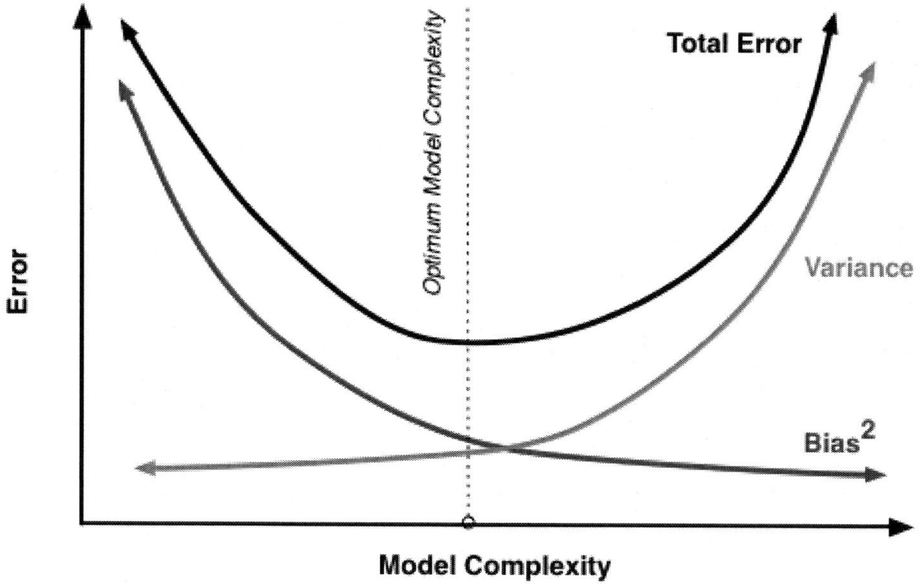

To find the best balance between bias and variance, different techniques can be used such as cross-validation, ensemble methods and regularization.

Regularization is a technique used to reduce the variance of a model by adding a penalty term to the cost function, which helps to prevent overfitting.

Cross-validation is a technique used to estimate the performance of a model on unseen data. By splitting the data into training and test sets, it is possible to estimate the performance of a model on unseen data, which helps to identify overfitting.

Ensemble methods are a set of techniques that combine the predictions of multiple models to improve overall performance. Ensemble methods are very powerful, as they can reduce both bias and variance.

In summary, understanding the bias-variance trade-off is essential for building good machine learning models. It is important to find a balance between bias and variance by adjusting model complexity and using techniques such as cross-validation, ensemble methods and regularization.

7.3 OVERFITTING AND UNDERFITTING

Overfitting and underfitting are two common problems in machine learning. These problems occur when a model is too complex or too simple for the data it is trying to fit, resulting in poor performance on new, unseen data. In this article, we will discuss overfitting and underfitting in detail and provide examples to help understand these concepts.

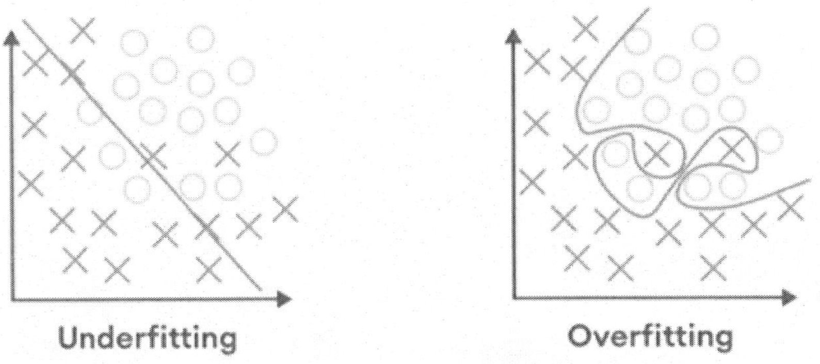

Overfitting

Overfitting occurs when a model is too complex and is able to fit the training data perfectly but performs poorly on new, unseen data. This happens when the model learns the noise in the data rather than the underlying pattern. As a result, the model becomes too specific to the training data and is unable to generalize to new data.

Example

Suppose we have a dataset that contains the age and height of a group of people. We want to build a model that predicts the height of a person given their age. We decide to use a polynomial regression model with a degree of 20 to fit the data.

```
import numpy as np
import matplotlib.pyplot as plt

# Generate random data
np.random.seed(42)
X = np.random.rand(50, 1)
y = X ** 2 + np.random.randn(50, 1) * 0.1

# Fit polynomial regression model
from sklearn.preprocessing import PolynomialFeatures
from sklearn.linear_model import LinearRegression
from sklearn.metrics import mean_squared_error

degrees = [1, 20]

for degree in degrees:
    poly_features = PolynomialFeatures(degree=degree,
include_bias=False)
    X_poly = poly_features.fit_transform(X)
    lin_reg = LinearRegression()
    lin_reg.fit(X_poly, y)
    y_pred = lin_reg.predict(X_poly)

    # Plot data and model predictions
    plt.scatter(X, y)
    plt.plot(X, y_pred, label='degree={}'.format(degree))
    plt.legend()
    plt.show()

    # Calculate mean squared error
    mse = mean_squared_error(y, y_pred)
    print('Degree {}: MSE = {}'.format(degree, mse))
```

The Output will look like this:

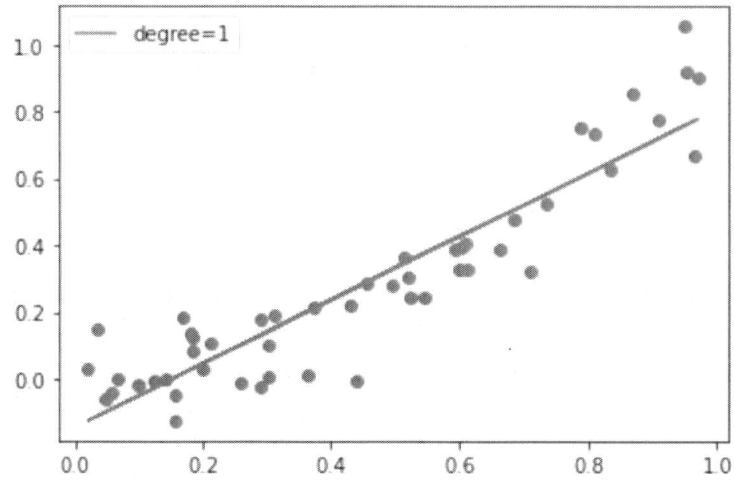


```
Degree 1: MSE = 0.013559152877012677
```

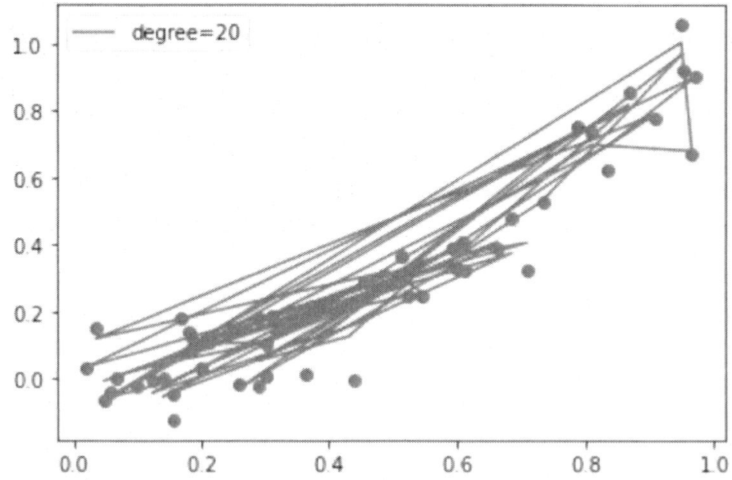


```
Degree 20: MSE = 0.004045051566147512
```

In this example, we generate random data with a quadratic relationship between age and height. We fit a polynomial regression model with degrees of 1 and 20 and plot the model predictions for each degree. As we can see from the plots, the model with a degree of 20 fits the training data perfectly, but is too complex and does not generalize well to new data. This is evident from the high mean squared error (MSE) value of the model with a degree of 20, which is much higher than the MSE value of the model with a degree of 1.

Underfitting

Underfitting occurs when a model is too simple and is unable to capture the underlying pattern in the data. This results in poor performance on both the training data and new, unseen data. Underfitting occurs when the model is not complex enough to capture the relationship between the features and the target variable.

Example

Suppose we have a dataset that contains the age and salary of a group of people. We want to build a model that predicts the salary of a person given their age. We decide to use a linear regression model to fit the data.

```
import numpy as np
import matplotlib.pyplot as plt

# Generate random data
np.random.seed(42)
X = np.random.rand(50, 1) * 10
y = X * 1000 + np.random.randn(50, 1) * 2000

# Fit linear regression model
from sklearn.linear_model import LinearRegression
from sklearn.metrics import mean_squared_error

lin_reg = LinearRegression()
```

```
lin_reg.fit(X, y)
y_pred = lin_reg.predict(X)

# Plot data and regression line
plt.scatter(X, y)
plt.plot(X, y_pred, color='red')
plt.title('Underfitting Example')
plt.xlabel('X')
plt.ylabel('y')
plt.show()

# Compute mean squared error
mse = mean_squared_error(y, y_pred)
print(f"Mean Squared Error: {mse}")
```

The output will look like this:

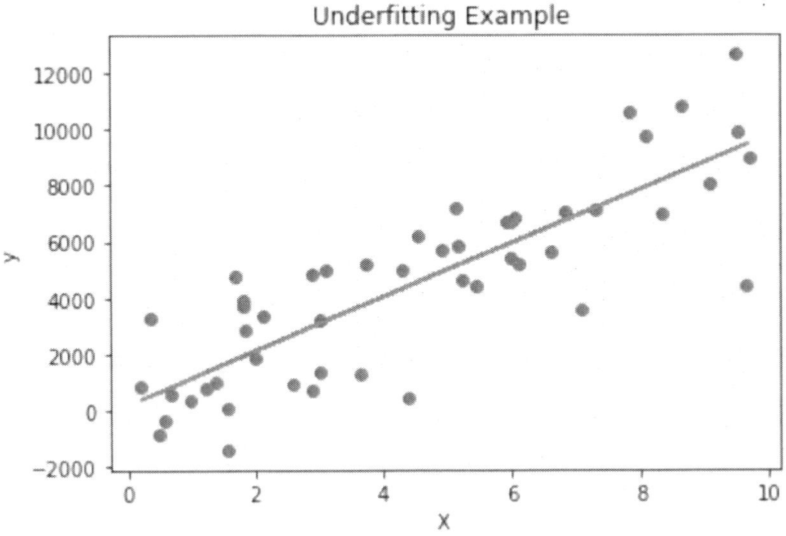


```
Mean Squared Error: 3292284.5749947513
```

The code first generates random data using the **numpy** library's **rand()** function. The data is then used to train a linear regression model using the **LinearRegression()** class from the **sklearn.linear_model** module. The model is then used to predict the target variable (**y**) and the predictions are stored in **y_pred**.

To visualize the data and regression line, **matplotlib** library is used to create a scatter plot of the data points and the predicted values by the model. The resulting plot shows that the linear regression line does not fit the data points well, indicating underfitting.

Finally, the mean squared error (MSE) is computed using the **mean_squared_error()** function from the **sklearn.metrics** module. The MSE value provides a measure of how well the model fits the data. In this case, the large value of MSE confirms that the model is not fitting the data well, further indicating underfitting.

7.4 SPLITTING THE DATA INTO TRAINING AND TESTING SETS

Splitting the data into training and testing sets is a fundamental technique in machine learning that is used to evaluate the performance of a model. The basic idea is to divide the available data into two parts: a training set and a testing set. The model is trained on the training set and its performance is evaluated on the testing set. This allows us to get an estimate of how well the model will perform on unseen data, which is critical for determining the model's ability to generalize to new data.

There are several ways to split the data into training and testing sets, with the most popular being:

Simple random sampling

Simple random sampling is a method for splitting data into training and testing sets, where the data is randomly split into two sets with a fixed ratio. This method is often used when the data is large, and the goal is to have a representative sample of the data for training and testing.

The process of simple random sampling is straightforward:

1. First, the data is shuffled randomly to remove any ordering or patterns.
2. Then, a fixed ratio is chosen for the split (e.g. 80% for training and 20% for testing).
3. Next, a random sample of the data is selected according to the chosen ratio, and this sample is used as the training set.
4. The remaining data is used as the testing set.

Simple random sampling is a simple and straightforward method for splitting data and it is easy to implement. However, it may not be suitable for datasets with imbalanced class distribution or small datasets. In these cases, stratified sampling could be a better choice.

It's important to keep in mind that when using simple random sampling, the training set and testing set may not be representative of the entire dataset. Therefore, it's important to shuffle the data randomly before splitting, to ensure that the samples are representative of the entire dataset.

An example of simple random sampling can be seen when creating a machine learning model to classify emails as spam or not spam. Suppose we have a dataset of 10,000 emails, where 7,000 are not spam and 3,000 are spam. We want to split the data into a training set and a testing set.

1. First, we shuffle the data randomly to remove any ordering or patterns.
2. Next, we choose a fixed ratio for the split, let's say 80% for training and 20% for testing.
3. We randomly select 8,000 emails (80% of the total emails) as the training set and 2,000 emails (20% of the total emails) as the testing set.
4. We use the training set to train our machine learning model and use the testing set to evaluate its performance.

It's important to note that in this example, the split ratio of 80% for training and 20% for testing is arbitrary, and the ratio can be adjusted to better suit the specific needs of the project.

In this example, we used simple random sampling to split the data into a training and testing set. We shuffled the data randomly to remove any ordering or patterns, and we chose a fixed ratio of 80% for training and 20% for testing. By using this method, we can use the training set to train our machine learning model and use the testing set to evaluate its performance. This method is simple and straightforward and it's easy to implement, but it may not be suitable for datasets with imbalanced class distribution or small datasets.

Here is an example of how simple random sampling can be implemented using Python's **train_test_split** function from the **sklearn.model_selection** module:

```
import numpy as np
from sklearn.model_selection import train_test_split

# Generate a random dataset with 1000 rows and 5 columns
X = np.random.rand(1000, 5)

# Generate a random target variable with 1000 rows
y = np.random.rand(1000)

# Split the dataset into training and testing sets
X_train, X_test, y_train, y_test = train_test_split(X, y,
test_size=0.3, random_state=42)

# Print the number of rows in the training and testing sets
print("Number of rows in the training set:", len(X_train))
print("Number of rows in the testing set:", len(X_test))
```

```
Number of rows in the training set: 700
Number of rows in the testing set: 300
```

In this example, we start by importing the necessary modules, including **numpy** and **sklearn.model_selection**. We then generate a random dataset with 1000 rows and 5 columns, and a random target variable with 1000 rows.

Next, we use the **train_test_split** function to split the dataset into training and testing sets. We specify that we want to use 30% of the data for testing by setting **test_size=0.3**, and we set the random seed to 42 using **random_state=42** to ensure that the split is reproducible.

Finally, we print the number of rows in the training and testing sets to verify that the split was performed correctly.

Simple random sampling is a commonly used technique in machine learning for splitting a dataset into training and testing sets. It involves randomly selecting a subset of the data to use for testing, while the remaining data is used for training.

This ensures that the testing set is representative of the entire dataset and can be used to evaluate the performance of a machine learning model. The **train_test_split** function in **sklearn.model_selection** makes it easy to perform simple random sampling in Python.

It's important to shuffle the data randomly before splitting to ensure that the samples are representative of the entire dataset.

Stratified sampling

Stratified sampling is a method for splitting data into training and testing sets, where the data is split into two sets in a way that **preserves the proportion of the target variable in both sets.** This method is useful when the data is imbalanced, meaning that the target variable has a disproportionate distribution among the different classes.

The process of stratified sampling is as follows:

1. First, the data is divided into different strata based on the value of the target variable.
2. Next, a fixed ratio is chosen for the split (e.g. 80% for training and 20% for testing).
3. A random sample of the data is then selected from each stratum according to the chosen ratio, and these samples are combined to form the training set.
4. The remaining data is used as the testing set.

Stratified sampling is useful when the data is imbalanced. With simple random sampling, the training set and testing set may not have the same proportion of the target variable as the entire dataset. This can lead to bias in the model and inaccurate predictions. By using stratified sampling, we ensure that the training set and testing set have the same proportion of the target variable as the entire dataset.

It's important to keep in mind that when using stratified sampling, the sample size may not be large enough to represent the entire dataset. Therefore, it's important to use a large enough sample size to ensure that the samples are representative of the entire dataset.

An example of stratified sampling can be seen when creating a machine learning model to classify credit card applications as approved or denied. Suppose we have a dataset of 10,000 applications, where 7,000 are approved and 3,000 are denied. We want to split the data into a training set and a testing set.

1. First, we divide the data into different strata based on the value of the target variable, in this case, approved or denied.

2. Next, we choose a fixed ratio for the split, let's say 80% for training and 20% for testing.

3. We randomly select 5,600 approved applications (80% of the total approved applications) and 2,400 denied applications (80% of the total denied applications) as the training set.

4. We use the remaining 1,400 approved applications (20% of the total approved applications) and 600 denied applications (20% of the total denied applications) as the testing set.

5. We use the training set to train our machine learning model and use the testing set to evaluate its performance.

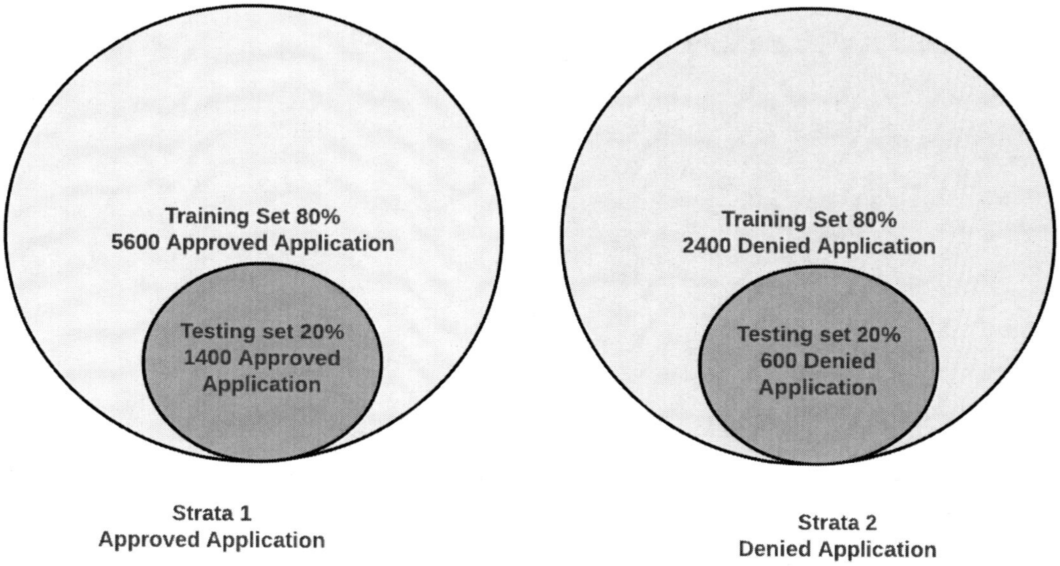

In this example, we used stratified sampling to split the data into a training and testing set. We divided the data into different strata based on the value of the target variable, approved or denied, and we chose a fixed ratio of 80% for training and 20% for testing. By using this method, we can ensure that the training set and testing set have the same proportion of the target variable as the entire dataset. This method is useful when the data is imbalanced, as it ensures that the training set and testing set have the same proportion of the target variable as the entire dataset.

It's important to note that in this example, the split ratio of 80% for training and 20% for testing is arbitrary, and the ratio can be adjusted to better suit the specific needs of the project.

Here is an example of how stratified sampling can be implemented using Python's **StratifiedShuffleSplit** class from the **sklearn.model_selection** module:

```
import numpy as np
from sklearn.datasets import make_classification
from sklearn.model_selection import StratifiedShuffleSplit

# Create a random dataset with 1000 samples and 5 features
X, y = make_classification(n_samples=1000, n_features=5,
random_state=42)

# Create an instance of StratifiedShuffleSplit with 5 splits and a
test size of 0.2
strat_split = StratifiedShuffleSplit(n_splits=5, test_size=0.2,
random_state=42)

# Iterate over each split and print the train and test indices
for train_index, test_index in strat_split.split(X, y):
    print("TRAIN:", train_index)
    print("TEST:", test_index)
    X_train, X_test = X[train_index], X[test_index]
    y_train, y_test = y[train_index], y[test_index]
```

In this example, we first create a random dataset with 1000 samples and 5 features using the **make_classification** function from scikit-learn. We then create an instance of **StratifiedShuffleSplit** with 5 splits and a test size of 0.2. We then iterate over each split and print the train and test indices.

The **split** method of **StratifiedShuffleSplit** takes in the feature matrix **X** and the target vector **y**. It returns an iterator that generates indices for the train and test sets for each split. We then use these indices to extract the corresponding subsets of the feature matrix **X** and target vector **y** for the train and test sets.

Stratified sampling is useful when dealing with imbalanced datasets where the number of samples in each class is not equal. It ensures that the test set contains representative samples from each class, which is important for evaluating the performance of a classifier on new, unseen data.

K-fold cross-validation

K-fold cross-validation is a method for evaluating the performance of a machine learning model by dividing the data into k folds (or subsets) and training the model k times, each time using a different fold as the testing set and the remaining k-1 folds as the training set. The performance of the model is then averaged across all k iterations to give a more robust estimate of its performance.

The process of k-fold cross-validation is as follows:

1. First, the data is divided into k equally sized folds.
2. Next, the model is trained k times, each time using a different fold as the testing set and the remaining k-1 folds as the training set.
3. The performance of the model is evaluated on the testing set and recorded.
4. The performance scores from all k iterations are then averaged to give a more robust estimate of the model's performance.

K-fold cross-validation is a commonly used method for evaluating the performance of machine learning models. It is particularly useful when the data is limited and we want to use as much of it as possible for training while still having a reliable estimate of the model's performance. It is also useful when the data has a high variance or when the model's performance is sensitive to the specific training and testing sets.

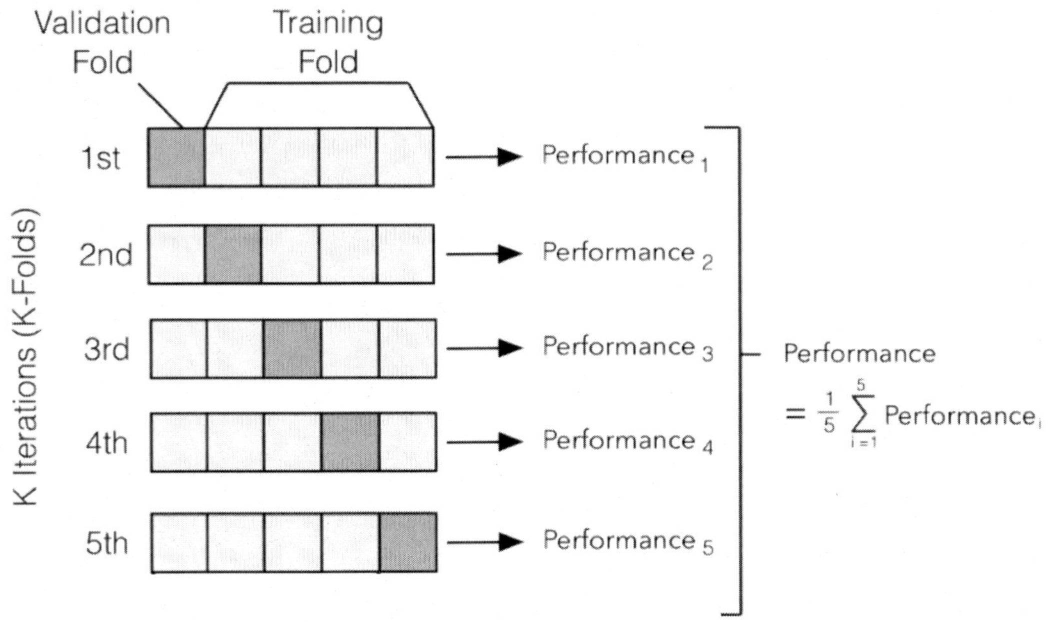

It's important to keep in mind that when using k-fold cross-validation, the sample size may not be large enough to represent the entire dataset. Therefore, it's important to use a large enough sample size to ensure that the samples are representative of the entire dataset.

An example of k-fold cross-validation can be seen when creating a machine learning model to classify customers as high-income or low-income. Suppose we have a dataset of 1,0000 customers and we want to evaluate the performance of our model.

1. First, we divide the data into 10 equally sized folds (k=10). Each fold contains 1000 customers.

2. Next, we train the model 10 times, each time using a different fold as the testing set and the remaining 9 folds as the training set.

3. We evaluate the performance of the model on the testing set using an appropriate metric such as accuracy, precision, recall or F1-score.

4. We record the performance scores from each iteration.

5. Finally, we average the performance scores to give a more robust estimate of the model's performance.

For example, if in the first iteration the model's accuracy is 90% on the testing set and 85% on the training set. In the second iteration, the model's accuracy is 92% on the testing set and 87% on the training set. After 10 iterations, we average all the accuracy scores of the model on the testing set, which will give us the overall performance of the model.

It's important to note that k-fold cross-validation can also be used with other evaluation metrics such as precision, recall, or F1-score. It's important to use the appropriate evaluation metric based on the problem at hand.

In this example, we used k-fold cross-validation to evaluate the performance of a machine learning model. We divided the data into 10 equally sized folds, trained the model 10 times, each time using a different fold as the testing set and evaluated its performance. We then averaged the performance scores across all iterations to give a more robust estimate of the model's performance.

Here is an example of how k-fold cross-validation can be implemented using Python's **KFold** class from the **sklearn.model_selection** module:

```
import pandas as pd
from sklearn.model_selection import KFold
from sklearn.linear_model import LogisticRegression
from sklearn.metrics import *

# Load data
data = pd.read_csv('customer.csv')
```

```
# Define features and target
X = data.drop(columns=['Segmentation', 'Profession', 'ID'], axis=1)
y = data['Segmentation']

# Create KFold object
kf = KFold(n_splits=5, shuffle=True, random_state=42)

# Initialize model
model = LogisticRegression()

# Create empty lists to store results
accuracy_scores = []
precision_scores = []
recall_scores = []
f1_scores = []

# Perform K-fold cross-validation
for train_index, test_index in kf.split(X):
    # Split the data into train and test sets
    X_train, X_test = X.iloc[train_index], X.iloc[test_index]
    y_train, y_test = y.iloc[train_index], y.iloc[test_index]

    # Train the model on the train set
    model.fit(X_train, y_train)

    # Test the model on the test set
    y_pred = model.predict(X_test)

    # Calculate evaluation metrics
    accuracy = model.score(X_test, y_test)
    precision = precision_score(y_test, y_pred, average='macro')
    recall = recall_score(y_test, y_pred, average='macro')
    f1 = f1_score(y_test, y_pred, average='macro')

    # Append results to the lists
    accuracy_scores.append(accuracy)
    precision_scores.append(precision)
    recall_scores.append(recall)
    f1_scores.append(f1)

# Print the average results
print("Average Accuracy:", sum(accuracy_scores)/len(accuracy_scores))
print("Average Precision:",
sum(precision_scores)/len(precision_scores))
print("Average Recall:", sum(recall_scores)/len(recall_scores))
print("Average F1-Score:", sum(f1_scores)/len(f1_scores))
```

The **KFold** class takes the input data (**X**) as argument, and the number of splits (**n_splits**) and random state (**random_state**) .The **shuffle** argument is used to shuffle the data before dividing it into folds.

In this example, we have a dataset of customer data with four different segments: A, B, C, and D. We want to train a logistic regression model to predict the customer segment based on the available features.

We first load the data into a pandas DataFrame and define the features and target variables. We then create a **KFold** object with five splits, set **shuffle** to **True** for random shuffling of the data before splitting, and **random_state** to 42 for reproducibility.

We then initialize the logistic regression model and create empty lists to store the evaluation metrics. We loop through each split of the data, train the model on the training set, and test it on the test set. We calculate the evaluation metrics for each split and append the results to the respective lists.

Finally, we print the average evaluation metrics over all splits of the data. This gives us an idea of how well the model is performing overall.

It is important to keep in mind that the way in which the data is split can have a significant impact on the performance of the model. If the data is not split properly, the model may be overfitting or underfitting, and the performance on unseen data may be poor.

It's also important to make sure that the training and testing sets are independent and identically distributed. This means that the training set and testing set should not overlap and should come from the same distribution. If the data is not independent and identically distributed, the model's performance on the testing set may not be a good estimate of its performance on unseen data.

7.5 HYPERPARAMETER TUNING

Hyperparameter tuning is the process of systematically searching for the best combination of hyperparameters in order to optimize the performance of a machine learning model. Hyperparameters are parameters that are not learned from data but are set by the user, such as the learning rate, the number of hidden layers, or the regularization strength. The optimal values of these parameters can greatly affect the performance of the model, and therefore it is important to tune them to achieve the best results.

There are several methods for tuning hyperparameters, including manual tuning, grid search, and random search.

Manual tuning

Manual tuning is the process of manually adjusting the hyperparameters and evaluating the performance of the model. It is the simplest method of hyperparameter tuning, as it involves manually adjusting the values of the hyperparameters and evaluating the model's performance.

The process of manual tuning involves the following steps:

1. Start with an initial set of hyperparameters.
2. Train the model using the initial hyperparameters.
3. Evaluate the model's performance on the validation set or using cross-validation.

4. Manually adjust one or more of the hyperparameters based on the evaluation results.

5. Repeat steps 2-4 until the model's performance reaches a satisfactory level or until the performance stops improving.

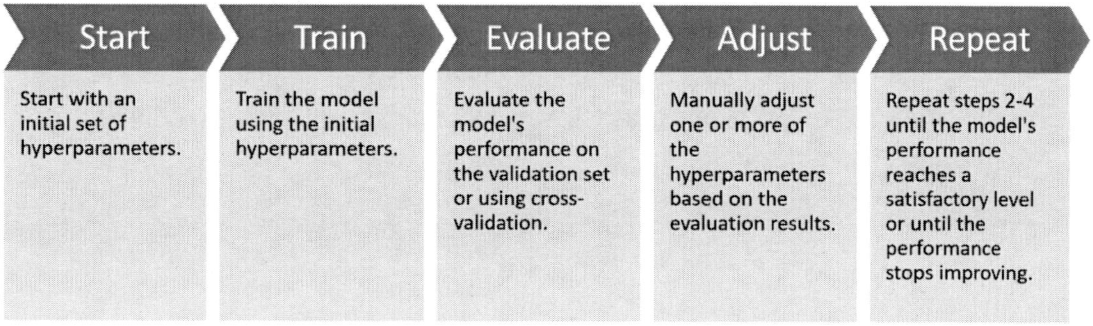

For example, if we are training a neural network, the initial set of hyperparameters might include the learning rate, the number of hidden layers, and the number of neurons in each layer. We would start with a small learning rate, a small number of hidden layers, and a small number of neurons in each layer. Then we would train the model and evaluate its performance. If the model's performance is poor, we would increase the learning rate and/or the number of hidden layers and/or the number of neurons in each layer. We would repeat this process until the model's performance reaches a satisfactory level or until the performance stops improving.

Manual tuning is simple to implement and can be effective when the number of hyperparameters is small and their possible values are limited. However, it can be time-consuming and may not always lead to the best results. It can also be impractical when the number of hyperparameters and their possible values is large.

In conclusion, manual tuning is a method for hyperparameter tuning that involves manually adjusting the values of the hyperparameters and evaluating the model's performance. It is simple to implement and can be effective when the number of hyperparameters is small and their possible values are limited. However, it can be time-consuming and may not always lead to the best results. It can also be impractical when the number of hyperparameters and their possible values is large.

Here is an example of how manual tuning can be implemented using Python's scikit-learn library:

```
import numpy as np
from sklearn.datasets import make_classification
from sklearn.model_selection import train_test_split
from sklearn.svm import SVC
from sklearn.metrics import accuracy_score

# Generate random dataset
np.random.seed(42)
X, y = make_classification(n_samples=1000, n_features=10, n_classes=2)

# Split the data into training and testing sets
X_train, X_test, y_train, y_test = train_test_split(X, y,
test_size=0.2, random_state=42)

# Initialize the SVM classifier
clf = SVC()

# Manually set hyperparameters
clf.kernel = 'rbf'  # radial basis function kernel
clf.C = 10  # regularization parameter
clf.gamma = 0.1  # kernel coefficient for rbf kernel

# Train the model on the training data
clf.fit(X_train, y_train)

# Predict the labels for the test data
y_pred = clf.predict(X_test)

# Evaluate the performance of the model
```

```
accuracy = accuracy_score(y_test, y_pred)
print('Accuracy:', accuracy)
```

The output will be:

```
Accuracy: 0.805
```

If we change the regularization parameter (clf.C) to 20 and kernal coefficient (clf.gamma) to 0.2, the accuracy will increase as below:

```
Accuracy: 0.825
```

In this example, we first generate a random dataset with 1000 samples and 10 features using the **make_classification()** function from the **sklearn.datasets** module. We then split the data into training and testing sets using the **train_test_split()** function from the **sklearn.model_selection** module, with a test size of 0.2 and a random state of 42 for reproducibility.

Next, we initialize the support vector machine (SVM) classifier using the **SVC()** function from the **sklearn.svm** module. We then manually set the hyperparameters of the SVM classifier: the kernel is set to 'rbf' (radial basis function kernel), the regularization parameter **C** is set to 10, and the kernel coefficient for the rbf kernel **gamma** is set to 0.1.

We then train the SVM classifier on the training data using the **fit()** method and predict the labels for the test data using the **predict()** method. Finally, we evaluate the performance of the model using the **accuracy_score()** function from the **sklearn.metrics** module.

By manually setting the hyperparameters, we can iteratively adjust the values until we achieve the desired level of performance. However, this process can be time-consuming and requires expert knowledge of the model and the dataset.

Automated hyperparameter tuning methods, such as grid search and random search, can help to streamline this process and find the optimal hyperparameters more efficiently.

Grid Search

Grid search is a method for systematically trying all possible combinations of hyperparameters within a predefined range. It is a computationally efficient method that can be used to find the optimal combination of hyperparameters for a machine learning model.

The process of grid search involves the following steps:

1. Define a search space for each hyperparameter.
2. Create a grid of all possible combinations of hyperparameters.
3. Train the model using each combination of hyperparameters in the grid.
4. Evaluate the model's performance on a validation set or using cross-validation.
5. Select the combination of hyperparameters that result in the best performance.

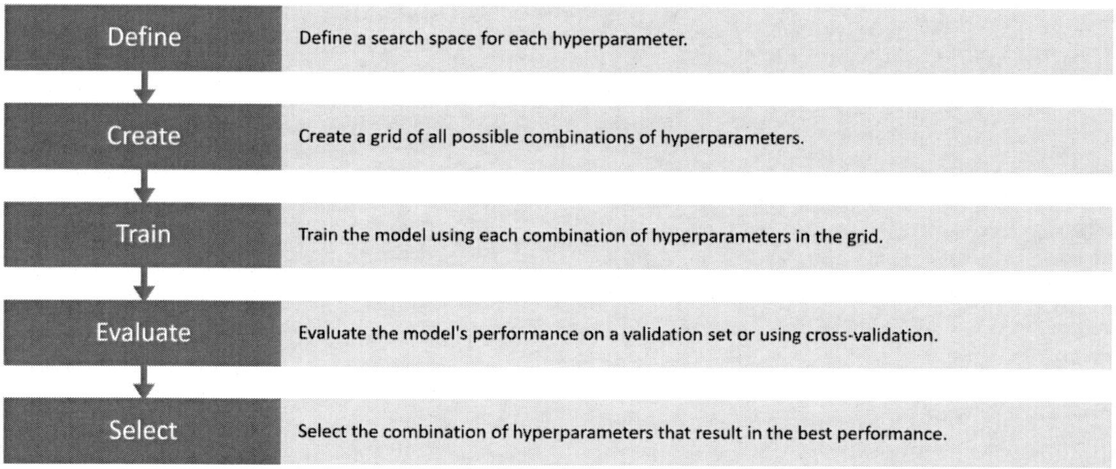

For example, if we are training a neural network, the search space for the learning rate might be [0.001, 0.01, 0.1], the search space for the number of hidden layers might be [1, 2, 3], and the search space for the number of neurons in each layer might be [50, 100, 150]. We would then create a grid of all possible combinations of these hyperparameters, train the model using each combination, and evaluate the model's performance on a validation set. The combination of hyperparameters that result in the best performance would be selected.

Python's scikit-learn library provides an easy-to-use implementation of grid search. The **GridSearchCV** class can be used to perform grid search on a given model, and it takes several important parameters:

- **estimator**: The model to be trained and evaluated.
- **param_grid**: A dictionary of hyperparameters to be searched over.
- **cv**: Number of cross-validation splits to use.
- **scoring**: A string or a callable to evaluate the predictions on the test set.
- **n_jobs**: The number of CPU cores used to perform the computation.

Here is an example of how grid search can be implemented using Python's scikit-learn library:

```
import numpy as np
from sklearn import datasets
from sklearn.ensemble import RandomForestClassifier
from sklearn.model_selection import GridSearchCV, train_test_split

# Generate random data
np.random.seed(42)
X, y = datasets.make_classification(n_samples=1000, n_features=10,
random_state=42)

# Split data into training and test sets
X_train, X_test, y_train, y_test = train_test_split(X, y,
test_size=0.3, random_state=42)

# Define the parameter grid to search over
param_grid = {
    'n_estimators': [10, 50, 100, 150, 200],
    'max_depth': [None, 5, 10, 15],
    'min_samples_split': [2, 5, 10],
    'min_samples_leaf': [1, 2, 4]
}

# Create the model to use for hyperparameter tuning
rfc = RandomForestClassifier(random_state=42)

# Perform grid search using 5-fold cross validation
grid_search = GridSearchCV(rfc, param_grid, cv=5, n_jobs=-1)

# Fit the grid search to the data
grid_search.fit(X_train, y_train)

# Print the best parameters and score
print(f"Best parameters: {grid_search.best_params_}")
print(f"Best score: {grid_search.best_score_}")

# Evaluate the model on the test set using the best parameters
best_rfc = grid_search.best_estimator_
y_pred = best_rfc.predict(X_test)
accuracy = np.mean(y_pred == y_test)
print(f"Accuracy on test set: {accuracy}")
```

In this example, we first generate a random dataset with **make_classification** from scikit-learn. We then split the data into training and test sets using **train_test_split**. Next, we define the parameter grid to search over using a dictionary where the keys are the hyperparameters we want to tune and the values are lists of values to try for each hyperparameter.

We create an instance of the model we want to use for hyperparameter tuning (in this case, a **RandomForestClassifier**) and pass it, along with the parameter grid and number of folds for cross-validation, to **GridSearchCV**. We then fit the grid search object to the training data.

After the grid search is complete, we print the best parameters and score using the **best_params_** and **best_score_** attributes of the grid search object. We then use the best estimator found by the grid search to make predictions on the test set, calculate the accuracy, and print it out.

Grid search is a useful technique for finding the best combination of hyperparameters for a machine learning model. By trying many different combinations and using cross-validation to evaluate their performance, we can find the set of hyperparameters that gives the best performance on our data.

Grid search is a powerful method to find the best set of hyperparameters for a given dataset and model. It is an efficient way of tuning the parameters of a model, and it can be used to find the optimal combination of hyperparameters for a machine learning model.

Random Search

Random search is an alternative method to grid search for hyperparameter tuning. Instead of trying every possible combination of hyperparameters, random search samples random combinations of hyperparameters from a predefined range. This allows for a more efficient exploration of the hyperparameter space, as it is not necessary to try every single combination.

The process of random search involves the following steps:

1. Define a search space for each hyperparameter.
2. Sample random combinations of hyperparameters from the search space.
3. Train the model using the sampled combination of hyperparameters.
4. Evaluate the model's performance on a validation set or using cross-validation.
5. Select the combination of hyperparameters that result in the best performance.

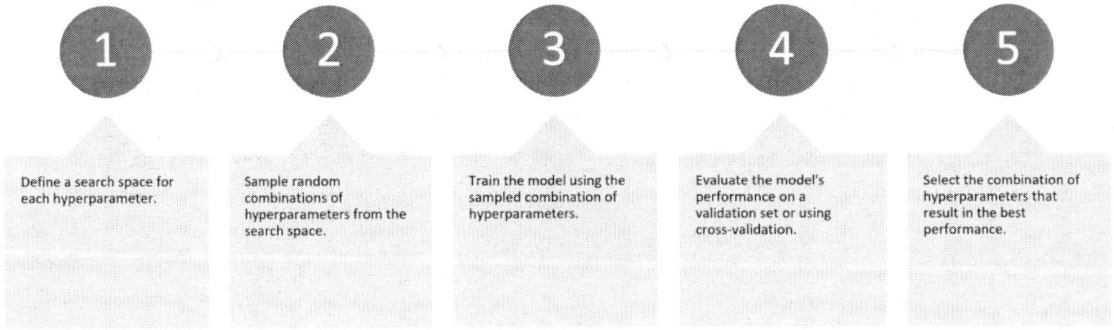

For example, if we are training a neural network, the search space for the learning rate might be [0.001, 0.01, 0.1], the search space for the number of hidden layers might be [1, 2, 3], and the search space for the number of neurons in each layer might be [50, 100, 150]. We would then sample random combinations of these hyperparameters, train the model using the sampled combination, and evaluate the model's performance on a validation set.

Python's scikit-learn library provides an easy-to-use implementation of random search. The **RandomizedSearchCV** class can be used to perform random search on a given model, and it takes several important parameters:

- **estimator**: The model to be trained and evaluated.
- **param_distributions**: A dictionary of hyperparameters to be searched over.
- **n_iter**: The number of random combinations of hyperparameters to be tried.
- **cv**: Number of cross-validation splits to use.
- **scoring**: A string or a callable to evaluate the predictions on the test set.
- **n_jobs**: The number of CPU cores used to perform the computation.

Here is an example of how random search can be implemented using Python's scikit-learn library:

```
import numpy as np
from sklearn.datasets import make_classification
from sklearn.model_selection import train_test_split,
RandomizedSearchCV
from sklearn.ensemble import RandomForestClassifier

# Generate random dataset
X, y = make_classification(n_samples=1000, n_features=10,
n_informative=5, n_classes=2, random_state=42)

# Split data into training and testing sets
```

```
X_train, X_test, y_train, y_test = train_test_split(X, y,
test_size=0.2, random_state=42)

# Define parameter grid
param_dist = {"n_estimators": [10, 50, 100, 200, 500],
              "max_depth": [2, 5, 10, 20, None],
              "min_samples_split": [2, 5, 10, 20],
              "min_samples_leaf": [1, 2, 4, 8],
              "max_features": ['sqrt', 'log2', None]}

# Define classifier
rfc = RandomForestClassifier()

# Create randomized search object
random_search = RandomizedSearchCV(rfc,
param_distributions=param_dist, n_iter=50, cv=5, random_state=42)

# Fit randomized search object to training data
random_search.fit(X_train, y_train)

# Print best hyperparameters
print("Best Hyperparameters:", random_search.best_params_)

# Evaluate best model on test data
y_pred = random_search.best_estimator_.predict(X_test)
accuracy = np.mean(y_pred == y_test)
print("Accuracy:", accuracy)
```

In this example, we first generate a random dataset of 1000 samples with 10 features and 2 classes using the make_classification function from scikit-learn. We then split the data into training and testing sets with a test size of 0.2.

Next, we define a parameter grid for the Random Forest Classifier (RFC) model. This grid contains a range of hyperparameters we want to tune, including the number of estimators, maximum depth, minimum samples to split, minimum samples per leaf, and maximum features.

We create an RFC classifier and a RandomizedSearchCV object. The latter is responsible for finding the best hyperparameters from the given parameter grid by performing cross-validation on the training set.

We fit the randomized search object to the training data and print the best hyperparameters found. Finally, we evaluate the best model on the test data by calculating the accuracy of its predictions.

Randomized search can be more efficient than grid search as it only samples a subset of the parameter grid, and is therefore useful when the hyperparameter search space is large. By using random sampling instead of exhaustive search, it can also avoid getting stuck in local optima.

Bayesian Optimization

Bayesian optimization is a method for hyperparameter tuning that uses Bayesian principles to model the function that maps hyperparameters to the performance of a machine learning model. It is particularly useful for expensive optimization problems, such as those that involve training large neural networks.

The basic idea behind Bayesian optimization is to use a probabilistic model to represent the relationship between the hyperparameters and the performance of the model. This model is then used to guide the search for the optimal hyperparameters. The model is updated after each iteration with the new data obtained from evaluating the model with different hyperparameters.

The process of Bayesian optimization involves the following steps:

1. Define a probabilistic model that maps hyperparameters to the performance of the model.

2. Sample the next set of hyperparameters to try by optimizing the acquisition function.

3. Train the model using the sampled hyperparameters.

4. Evaluate the model's performance on a validation set or using cross-validation.

5. Update the probabilistic model with the new data.

6. Repeat steps 2-5 until a stopping criterion is met.

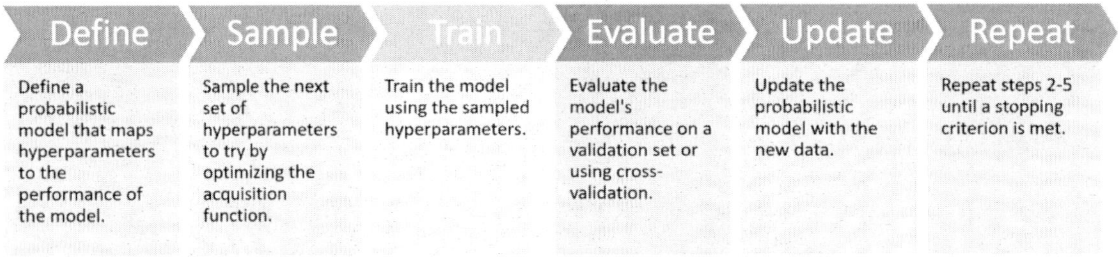

There are several different probabilistic models that can be used for Bayesian optimization, such as Gaussian processes and random forests. The choice of probabilistic model will depend on the specific problem and the available computational resources.

The acquisition function is used to guide the search for the next set of hyperparameters to try. It balances the exploration of the hyperparameter space with the exploitation of the current knowledge about the function that maps hyperparameters to the performance of the model. Commonly used acquisition functions include expected improvement, upper confidence bound, and probability of improvement.

You may need to install bayesian-optimization (if you already haven't installed) to run the below example. To install bayesian-optimization, you can run the below command:

pip install bayesian-optimization

Here's an example of using Bayesian Optimization with a random dataset:

```
import numpy as np
from sklearn.model_selection import cross_val_score
from bayes_opt import BayesianOptimization
from sklearn.ensemble import RandomForestRegressor

# Generate random data
np.random.seed(42)
X = np.random.rand(100, 5)
y = np.random.rand(100)

# Define the model to be optimized
def rf_cv(n_estimators, min_samples_split, max_features, data,
targets):
    estimator = RandomForestRegressor(
        n_estimators=int(n_estimators),
        min_samples_split=int(min_samples_split),
        max_features=min(max_features, 0.999),
        random_state=42
    )
    cval = cross_val_score(estimator, data, targets,
scoring='neg_mean_squared_error', cv=4)
    return cval.mean()

# Set the parameter bounds for Bayesian Optimization
pbounds = {
    'n_estimators': (10, 250),
    'min_samples_split': (2, 25),
    'max_features': (0.1, 0.999)
}

# Run the Bayesian Optimization process
optimizer = BayesianOptimization(
    f=rf_cv,
    pbounds=pbounds,
```

```
    random_state=42,
)
optimizer.maximize(init_points=10, n_iter=20)

# Print the best parameters found by the optimization process
print(optimizer.max)
```

In this example, we are generating a random dataset of 100 observations and 5 features, along with a corresponding target variable. We then define a Random Forest Regression model to be optimized using Bayesian Optimization.

We set the parameter bounds for the optimization process, which are the ranges within which the optimizer will search for the optimal set of hyperparameters. We then define a function, **rf_cv**, which takes the hyperparameters as inputs, fits a Random Forest Regression model with those hyperparameters, and returns the negative mean squared error obtained via 4-fold cross-validation.

We then run the Bayesian Optimization process using the **BayesianOptimization** class from the **bayes_opt** module. We provide the **rf_cv** function as the objective function to be optimized, along with the parameter bounds and a random seed. We then call the **maximize** method, which performs the optimization process with 10 initial random points and 20 iterations of the optimization algorithm.

Finally, we print out the best set of hyperparameters found by the optimization process.

If you don't want to use **bayes_opt** module and you want to use your own data and model to test Bayesian Optimization, continue to read below:

Python's scikit-learn library provides an easy-to-use implementation of Bayesian optimization through the **BayesSearchCV** class. The **BayesSearchCV** class can be used to perform Bayesian optimization on a given model, and it takes several important parameters:

- **estimator**: The model to be trained and evaluated.
- **search_spaces**: A dictionary of hyperparameters to be searched over.
- **n_iter**: The number of iterations to run the optimization.
- **cv**: Number of cross-validation splits to use.
- **scoring**: A string or a callable to evaluate the predictions on the test set.
- **n_jobs**: The number of CPU cores used to perform the computation.

Here is an example of how Bayesian optimization can be implemented using Python's scikit-learn library:

```
from sklearn.model_selection import BayesSearchCV
from sklearn.metrics import accuracy_score

# load the data
X, y = load_your_data()

# define the search space
param_space = {'learning_rate': (0.001, 0.1),
               'num_hidden_layers': (1, 4),
               'num_neurons': (50, 150)}

# create the Bayesian optimization object
clf = YourModel()
bayes_search = BayesSearchCV(clf, param_space, n_iter=10, cv=5,
scoring='accuracy', n_jobs=-1)

# perform Bayesian optimization
bayes_search.fit(X, y)
```

```
# print the best parameters and the best score
print("Best parameters:", bayes_search.best_params_)
print("Best score:", bayes_search.best_score_)

# retrain the model with the best parameters
best_model = bayes_search.best_estimator_
best_model.fit(X, y)

# evaluate the model on the test set
y_test = load_your_test_data()
y_pred = best_model.predict(y_test)
score = accuracy_score(y_test, y_pred)
print("Accuracy on test set:", score)
```

In this example, the data is loaded and the search space for the hyperparameters is defined. Then, the **BayesSearchCV** object is created, specifying the model, the search space, the number of iterations to run the optimization, the number of cross-validation splits, the scoring metric and the number of CPU cores used to perform the computation. After that, the **BayesSearchCV** object's fit method is called passing in the feature and target variable, this will perform the Bayesian optimization and it will return the best combination of hyperparameters. Finally, the best model is retrained using the best parameters and its performance is evaluated on the test set using the accuracy_score function and the accuracy score is printed out. It is important to note that the **load_your_data()** and **YourModel()** should be replaced by the actual code for loading the data and initializing the model used for the specific task.

Bayesian optimization is particularly useful for expensive optimization problems, such as those that involve training large neural networks. It is an efficient way of tuning the parameters of a model and it can be used to find the optimal combination of hyperparameters for a machine learning model.

In conclusion, Hyperparameter tuning is the process of systematically searching for the best combination of hyperparameters in order to optimize the performance of a machine learning model. There are several methods for tuning hyperparameters, including manual tuning, grid search, random search and Bayesian optimization. The choice of method will depend on the specific problem and the number of hyperparameters and their possible values.

7.6 MODEL INTERPRETABILITY

Model interpretability refers to the ability to understand and explain the decisions and predictions made by a machine learning model. It is an important aspect of machine learning as it allows practitioners to understand how a model is making its decisions, identify any potential biases, and make adjustments as necessary.

It's important to note that model interpretability and model performance are often trade-offs. Complex models such as deep neural networks can have better performance but are harder to interpret. Therefore, it's important to strike a balance between model interpretability and performance depending on the use case.

Model interpretability is an important aspect of machine learning that allows practitioners to understand and explain the decisions and predictions made by a model. Techniques such as feature importance analysis, model visualization, simplifying the model, and model-agnostic interpretability can be used to improve the interpretability of a model. However, it's important to strike a balance between model interpretability and performance depending on the use case.

There are several techniques that can be used to improve the interpretability of a machine learning model. Let's discuss some of the major techniques in the following sections.

7.7 FEATURE IMPORTANCE ANALYSIS

Feature importance analysis is a technique used to identify the most important features or variables that are contributing to the predictions made by a machine learning model. This can be useful for understanding how a model is making its decisions, identifying potential biases, and making adjustments as necessary.

There are several methods that can be used to perform feature importance analysis, including:

Permutation Importance:

Permutation Importance is a method used to determine the importance of each feature in a machine learning model. It is a simple and computationally efficient way of measuring feature importance. The basic idea behind permutation importance is to measure the change in model performance by randomly shuffling the values of a single feature, and then comparing the model's performance on the shuffled dataset to its performance on the original dataset. The features that result in the largest decrease in performance are considered the most important.

The process of permutation importance can be described in the following steps:

1. Fit the original model on the training dataset.
2. For each feature in the dataset, create a copy of the dataset with the values of that feature shuffled.
3. Retrain the model on the shuffled dataset and calculate the performance metric.

4. Compare the performance metric of the model on the shuffled dataset to the performance metric on the original dataset.

5. Repeat steps 2 to 4 for all features in the dataset.

6. The feature that causes the largest decrease in performance when shuffled is considered the most important feature.

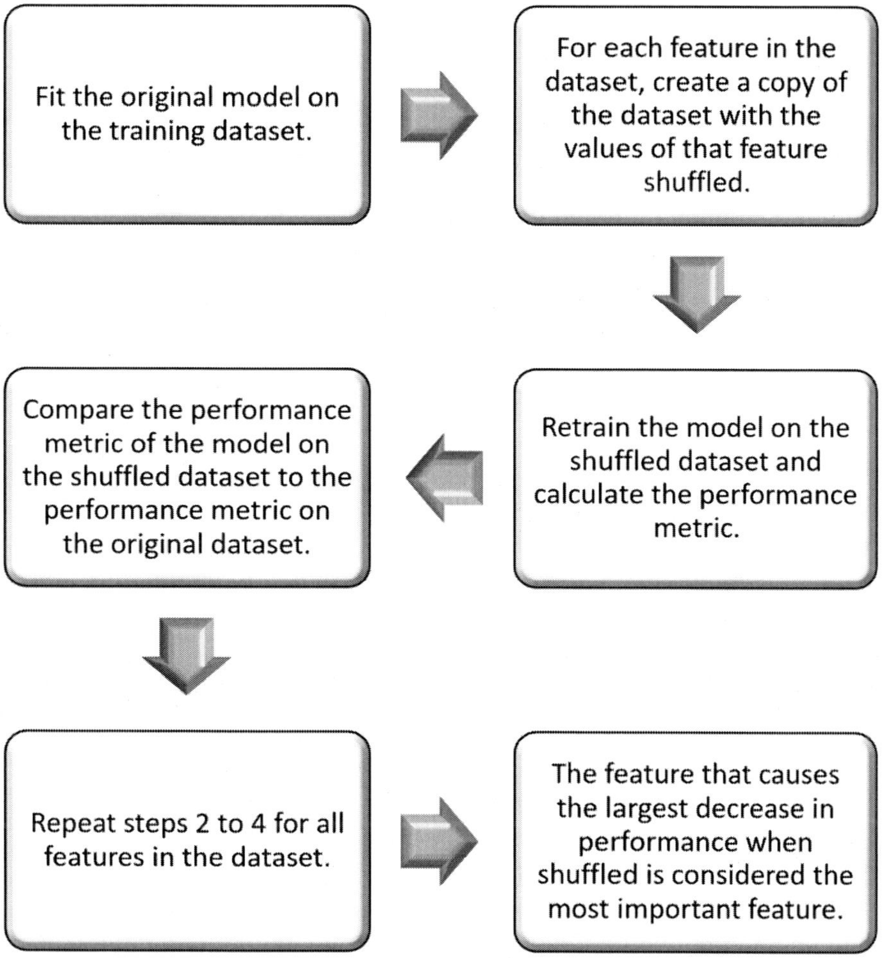

It's important to note that permutation importance is model-dependent, meaning that the results will vary depending on the type of model being used. Additionally, permutation importance should be interpreted with caution as it may not always reflect the true underlying relationship between a feature and the target variable.

Here's an example of how to calculate permutation importance using the scikit-learn library in Python:

```
from sklearn.ensemble import RandomForestRegressor
from sklearn.metrics import mean_squared_error
from sklearn.datasets import make_regression
import numpy as np

# Generate a synthetic dataset for regression
X, y = make_regression(n_features=4, random_state=0)

# Fit a random forest model
rf = RandomForestRegressor(random_state=0)
rf.fit(X, y)

# Initialize an empty dictionary to store the permutation importances
perm_importances = {}

# Iterate over each feature
for feature in range(X.shape[1]):
    # Shuffle the values of the feature
    X_shuffled = X.copy()
    X_shuffled[:, feature] = np.random.permutation(X_shuffled[:,
feature])

    # Compute the model's performance on the shuffled dataset
    y_pred = rf.predict(X_shuffled)
    score = mean_squared_error(y, y_pred)

    # Store the feature's permutation importance
    perm_importances[feature] = score

# Print the permutation importances
print(perm_importances)
```

In this example, we first generate a synthetic dataset for regression using the **make_regression** function from scikit-learn. Then, we fit a random forest model on the dataset using the **RandomForestRegressor** class.

Next, we initialize an empty dictionary to store the permutation importances and iterate over each feature in the dataset. For each feature, we shuffle the values using the **np.random.permutation** function and compute the model's performance on the shuffled dataset using the **mean_squared_error** function. We then store the feature's permutation importance in the dictionary.

Finally, we print the permutation importances to see which features are the most important according to the model.

It's worth noting that this example is used for illustration purposes and the feature importance may change if the data changes or the model changed. Also, it's important to check the permutation importance for different models as the importance of a feature might be different for different models.

One real-life application of permutation importance could be in the field of finance, where a model is used to predict stock prices. In this case, permutation importance can be used to determine which factors are most important in determining stock prices. For example, the company's financial statements, economic indicators, and news articles are features that can be used to predict stock prices. By using permutation importance, the analyst can understand which of these factors have the most impact on stock prices, and thus, make better investment decisions.

Another example could be in healthcare where a model is used to predict patient outcomes. Here, permutation importance can be used to determine which patient characteristics are most important in predicting outcomes such as survival rate or length of stay in the hospital. By identifying the most important patient characteristics, doctors and healthcare professionals can make more informed decisions about patient treatment.

In conclusion, permutation importance is a versatile technique that can be applied in various domains such as finance, healthcare, and many other fields. It's a way to understand which factors have the most impact on a model's predictions which can be extremely useful in making informed decisions.

SHAP values (SHapley Additive exPlanations)

SHAP values, or SHapley Additive exPlanations, is a method for interpreting the output of any machine learning model. It is based on the concept of Shapley values from cooperative game theory, which provides a way to fairly distribute a value among a group of individuals. In the context of machine learning, SHAP values can be used to understand the contribution of each feature to the prediction of a specific instance.

The basic idea behind SHAP values is to estimate the contribution of each feature to the prediction of an instance by considering all possible coalitions of features. A coalition is a subset of features that can be used to make a prediction. The contribution of each feature is calculated by averaging its marginal contribution to all coalitions that include that feature.

SHAP values have several important properties:

- They are model-agnostic, meaning they can be used with any type of machine learning model.
- They provide a unified measure of feature importance that is consistent across all instances.

- They take into account the interaction of features and the dependencies between them.

To compute SHAP values, the following steps are followed:

A. First, a baseline value is established, which represents the expected prediction of the model when all features are set to their expected values.

B. Next, the contribution of each feature to the prediction of the instance is calculated by comparing the prediction of the model when that feature is set to its actual value to the baseline value.

C. Finally, the contributions of all features are combined to obtain the final SHAP value for the instance.

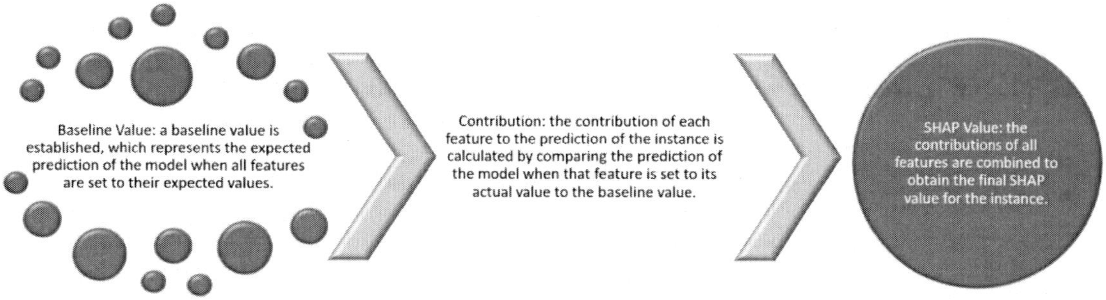

SHAP values can be visualized using a variety of techniques such as summary plots, dependence plots, and force plots. Summary plots provide a global view of feature importance by showing the average contribution of each feature across all instances. Dependence plots show the relationship between a feature and the prediction for a specific instance. Force plots provide an interactive visualization of the contributions of each feature to the prediction of a specific instance.

If shap is not installed on your device, you need to install shap by using the below command:

pip install shap

Here's an example of how to compute and visualize SHAP values using the scikit-learn library in Python:

```
import shap
from sklearn.ensemble import RandomForestRegressor
from sklearn.datasets import make_regression

# Generate a synthetic dataset for regression
X, y = make_regression(n_features=4, random_state=0)

# Fit a random forest model
rf = RandomForestRegressor(random_state=0)
rf.fit(X, y)

# Compute the SHAP values for the first instance in the dataset
explainer = shap.Explainer(rf, X[0])
shap_values = explainer.shap_values()

# Print the SHAP values
print(shap_values)

# Plot the summary plot
shap.summary_plot(shap_values)

# Plot the dependence plot
shap.dependence_plot("Feature 0", shap_values, X)

# Plot the force plot
shap.force_plot(explainer.expected_value, shap_values[0], X[0])
```

In this example, we first generate a synthetic dataset for regression using the **make_regression** function from scikit-learn. Then, we fit a random forest model on the dataset using the **RandomForestRegressor** class.

Next, we create an instance of the **Explainer** class from the shap library and pass the model and the first instance of the dataset to it. Then, we use the **shap_values()** method to compute the SHAP values for that instance.

Then, we use different plotting functions from the shap library to visualize the results. **summary_plot** plots the average contribution of each feature across all instances, **dependence_plot** plots the relationship between a feature and the prediction for a specific instance and **force_plot** provides an interactive visualization of the contributions of each feature to the prediction of a specific instance.

It's worth noting that the above example is used for illustration purposes and the results may change if the data changes or the model changed. Also, it's important to check the SHAP values for different models as the importance of a feature might be different for different models.

One real-life application of SHAP values could be in the field of healthcare, where a model is used to predict patient outcomes such as survival rate or length of stay in the hospital. In this case, SHAP values can be used to understand the contribution of each patient characteristic to the prediction of a specific patient outcome.

For example, a model could be trained using patient data such as age, gender, medical history, laboratory results, and vital signs to predict the survival rate of a patient with a specific disease. By using SHAP values, the healthcare professionals can understand which patient characteristics have the most impact on the survival rate and make more informed decisions about patient treatment.

Another example could be in finance, where a model is used to predict stock prices. In this case, SHAP values can be used to understand the contribution of each feature to the prediction of stock prices. For example, the company's financial statements, economic indicators, and news articles are features that can be used to predict stock prices. By using SHAP values, the analyst can understand which of these factors have the most impact on stock prices, and thus, make better investment decisions.

SHAP values can be applied in various domains such as finance, healthcare, and many other fields. It's a powerful tool that can be used to understand the contribution of each feature to the prediction of a specific instance, which can be extremely useful in making informed decisions.

SHAP values (SHapley Additive exPlanations) is a powerful tool for interpreting the output of any machine learning model. It is based on the concept of Shapley values from cooperative game theory and provides a unified measure of feature importance that is consistent across all instances. It takes into account the interaction of features and the dependencies between them. It can be visualized in various ways and can be used in any domain where machine learning models are used.

Partial dependence plots

Partial dependence plots (PDPs) are a popular technique used to understand the relationship between a feature and the prediction of a machine learning model. They provide a way to visualize the average effect of a feature on the prediction while holding all other features constant.

A PDP is a plot that shows the relationship between the value of a feature and the prediction of the model. The x-axis of the plot represents the values of the feature, and the y-axis represents the predicted value of the model. The plot is generated by fixing the values of all other features to their average values and then varying the value of the feature of interest. The plot shows the average prediction of the model for all instances that have the same value of the feature of interest.

PDPs are useful for understanding the relationship between a feature and the prediction of a model. They can help identify non-linear relationships between features and the prediction and can also help identify interactions between features. PDPs are also useful for identifying which features are important for the model's prediction and for understanding the relative importance of different features.

PDPs can be generated using the **plot_partial_dependence** function in the scikit-learn library. The function takes the following arguments:

- **estimator**: The fitted model object
- **X**: The feature dataset
- **features**: The feature or features for which the PDP is to be plotted
- **feature_names**: The names of the features
- **response_name**: The name of the response variable

Here is an example of how to generate a PDP using the scikit-learn library in Python:

```
from sklearn.ensemble import RandomForestRegressor

from sklearn.datasets import make_regression
from sklearn.inspection import plot_partial_dependence

# Generate a synthetic dataset for regression
X, y = make_regression(n_features=4, random_state=0)

# Fit a random forest model
rf = RandomForestRegressor(random_state=0)
rf.fit(X, y)

# Generate PDP for feature 0
```

```
plot_partial_dependence(rf, X, [0], feature_names=['Feature 0'])
```

```
<sklearn.inspection._plot.partial_dependence.PartialDependenceDisplay at 0x23b442df550>
```

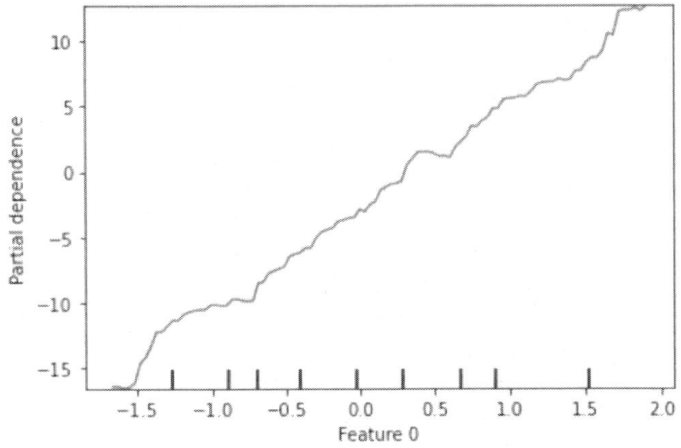

In this example, we first generate a synthetic dataset for regression using the **make_regression** function from scikit-learn. Then, we fit a random forest model on the dataset using the **RandomForestRegressor** class.

Next, we use the **plot_partial_dependence** function from the scikit-learn library to generate a PDP for feature 0. The function takes the fitted model, the feature dataset, the feature of interest and the feature names as input.

It's worth noting that this is a simple example, and in real-world applications, the data might be more complex, and the features might be correlated. Also, it's important to check the PDPs for different models as the relationship between a feature and the output might be different for different models.

LIME (Local Interpretable Model-agnostic Explanations)

LIME (Local Interpretable Model-agnostic Explanations) is a technique used to interpret the predictions of complex machine learning models. It is a model-agnostic approach, which means it can be applied to any type of model, regardless of its architecture or algorithm.

The idea behind LIME is to explain the predictions of a model by training a simple interpretable model on a small subset of the data, locally around the instance of interest. This approach allows us to understand how the model is making its predictions, even for instances where the global behavior of the model is not clear.

Here's how LIME works:

1. For an instance of interest, a perturbation of the data is generated by randomly sampling instances from the dataset and replacing some of the feature values of the instance of interest with those of the sampled instances.
2. A simple interpretable model (e.g. linear regression) is trained on the perturbed data.
3. The coefficients of the interpretable model are used as feature importances to explain the prediction of the instance of interest.

If lime is not installed on your device, you need to install lime by using the below command:

pip install lime

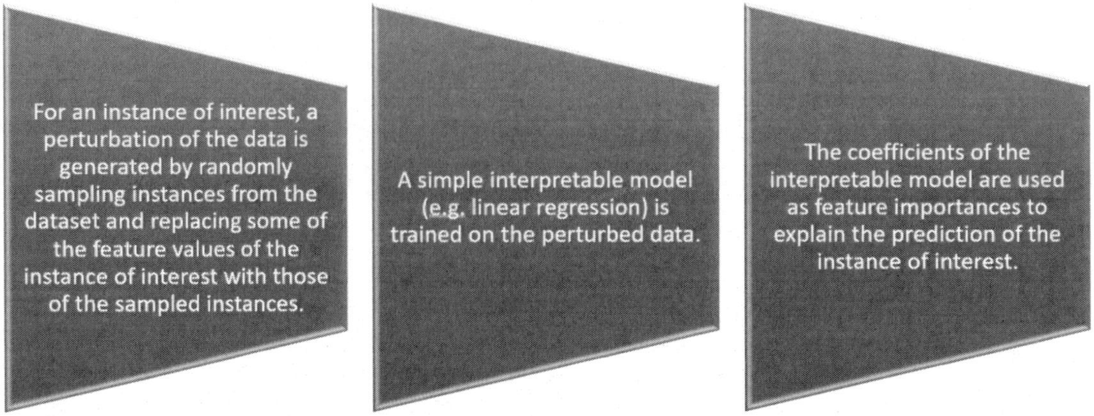

LIME can be implemented in Python using the **lime** library. Here is an example of how to use LIME to explain a prediction of a random forest model:

```
from lime import lime_tabular
from sklearn.ensemble import RandomForestClassifier
from sklearn.datasets import make_classification

# Generate a synthetic dataset for classification
X, y = make_classification(n_features=4, random_state=0)

# Fit a random forest model
rf = RandomForestClassifier(random_state=0)
rf.fit(X, y)

# Create an explainer object
explainer = lime_tabular.LimeTabularExplainer(X,
feature_names=['Feature 0', 'Feature 1', 'Feature 2', 'Feature 3'],
class_names=['Class 0', 'Class 1'])

# Explain a prediction
instance = X[0]
exp = explainer.explain_instance(instance, rf.predict_proba,
num_features=4)
```

In this example, we first generate a synthetic dataset for classification using the **make_classification** function from scikit-learn. Then, we fit a random forest model on the dataset using the **RandomForestClassifier** class.

Next, we create an explainer object using the **lime_tabular.LimeTabularExplainer** class. This class takes the feature dataset, feature names, and class names as input.

Finally, we use the **explain_instance** method of the explainer object to explain a prediction of the random forest model for the first instance in the dataset. The method takes the instance of interest, the prediction function and the number of features to use as input.

The result of the **explain_instance** method is an explanation object that contains the feature importances and the predicted class. The feature importances can be visualized using the **as_pyplot_figure** method of the explanation object.

One important thing to keep in mind when using LIME is that it can only explain the predictions of a model for a local region of the data. Therefore, it's not suitable for understanding the global behavior of the model or for identifying global patterns in the data. Additionally, LIME is computationally expensive, especially for large datasets and complex models. Therefore, it's important to use it judiciously and only when necessary.

LIME is a useful technique for interpreting the predictions of complex machine learning models. It allows us to understand how the model is making its predictions locally around an instance of interest, and it can be used to compare different models, identify important features and improve the interpretability of a model.

eli5 library

The ELI5 (Explain Like I'm 5) library is a Python library for explaining and interpreting machine learning models. It is built on top of the scikit-learn library and provides a simple and intuitive interface for understanding the predictions of complex models.

One of the main features of the ELI5 library is its ability to generate explanations for individual predictions in a human-readable format. This is achieved by using techniques such as LIME (Local Interpretable Model-agnostic Explanations) and SHAP (SHapley Additive exPlanations) to compute feature importances and generate explanations for the model's predictions.

The library also provides a number of other useful features such as the ability to visualize feature importances and explanations, support for different types of models and datasets, and the ability to debug and debug models.

If *eli5* is not installed on your device, you need to install eli5 by using the below command:

pip install eli5

Here's an example of how to use the ELI5 library to explain a prediction of a random forest model:

```
import eli5
from sklearn.ensemble import RandomForestClassifier
from sklearn.datasets import make_classification

# Generate a synthetic dataset for classification
X, y = make_classification(n_features=4, random_state=0)

# Fit a random forest model
```

```
rf = RandomForestClassifier(random_state=0)
rf.fit(X, y)

# Explain a prediction
instance = X[0]
exp = eli5.explain_prediction(rf, instance)
```

In this example, we first generate a synthetic dataset for classification using the **make_classification** function from scikit-learn. Then, we fit a random forest model on the dataset using the **RandomForestClassifier** class.

Next, we use the **explain_prediction** function of the ELI5 library to explain a prediction of the random forest model for the first instance in the dataset. The function takes the model and the instance of interest as input.

The result of the **explain_prediction** function is an explanation object that contains the feature importances, the predicted class and the explanation of the prediction in a human-readable format. The feature importances and the explanation can be visualized using the **show_weights** and **show_prediction** methods of the explanation object, respectively.

It's important to note that feature importance analysis is model-dependent, meaning that the results will vary depending on the type of model being used. Additionally, feature importance should be interpreted with caution as it may not always reflect the true underlying relationship between a feature and the target variable.

Feature importance analysis is a technique used to identify the most important features or variables that are contributing to the predictions made by a machine learning model. There are various methods such as Permutation Importance, SHAP values, Partial dependence plots, LIME, eli5 library that can be used to perform feature importance analysis. However, it's important to keep in mind that feature importance results can vary depending on the model and should be interpreted with caution.

7.8 MODEL VISUALIZATION

Model visualization is the process of creating visual representations of machine learning models to help understand and interpret their behavior. This is particularly useful for understanding complex models such as deep neural networks, which can be difficult to interpret based on their internal parameters alone. Model visualization can be used to gain insight into the model's structure, identify patterns and dependencies in the data, and understand how the model is making its predictions.

There are several techniques used for model visualization, including:

Activation maps

Activation maps, also known as feature maps, are a technique used to visualize the activations of individual neurons in a neural network. These maps provide a way to understand how the model is processing the input data, and can be used to gain insight into the model's behavior.

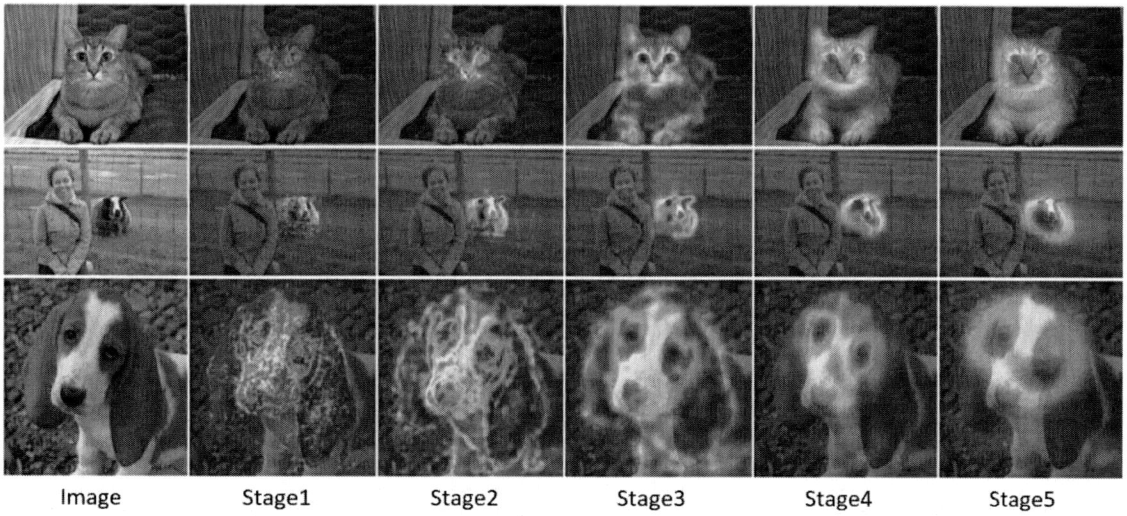

Image Source (https://mmcheng.net/layercam/)

Activation maps are typically created by passing an input image through the network and visualizing the output of a specific layer. For example, we can pass an image of a handwritten digit through a convolutional neural network (CNN) and visualize the output of the first convolutional layer. The output of this layer, also called feature maps, contains a set of filtered images, each representing a specific feature of the input image.

If *keras* is not installed on your device, you need to install keras by using the below command:

pip install keras

Here is an example of how to create an activation map using the Keras library:

```
from keras.models import Model

# Load a pre-trained model
model = keras.applications.VGG16(weights='imagenet',
include_top=False)

# Choose a specific input image
img = keras.preprocessing.image.load_img('image.jpg',
target_size=(224, 224))
x = keras.preprocessing.image.img_to_array(img)
x = np.expand_dims(x, axis=0)

# Pass the input image through the model
features = model.predict(x)

# Visualize the activations of the first convolutional layer
first_layer_activation = features[:, :, :, :]
plt.matshow(first_layer_activation[0, :, :, 0], cmap='viridis')
plt.show()
```

In this example, we first load a pre-trained VGG16 model and choose a specific input image. We then pass the input image through the model, and visualize the activations of the first convolutional layer using the **matshow** function from the **matplotlib** library. The resulting image shows the filtered images produced by the first convolutional layer, highlighting the features of the input image that the network has identified.

Activation maps are useful for understanding the features that a neural network is learning and how it processes the input data. This can help in identifying the problem areas in the model and fine-tuning the model accordingly. Activation maps can also be used for data augmentation, where you can use the feature maps to generate new images.

In conclusion, activation maps are a powerful technique for visualizing the activations of individual neurons in a neural network. They provide a way to understand how the model is processing the input data, and can be used to gain insight into the model's behavior. Activation maps are widely used for understanding the features learned by the model and fine-tuning the model accordingly.

Layer-wise relevance propagation (LRP)

Layer-wise relevance propagation (LRP) is a method for understanding and interpreting the predictions made by neural networks. It is based on the idea of propagating the relevance of the output predictions back through the network, layer by layer, to the input features. This helps to determine which input features were most important for the final prediction.

The basic idea behind LRP is that the relevance score of each output neuron is propagated back through the network, layer by layer, to the input neurons. This relevance score is calculated based on the contribution of each neuron to the final prediction. The relevance score is then used to highlight which input features were most important for the final prediction.

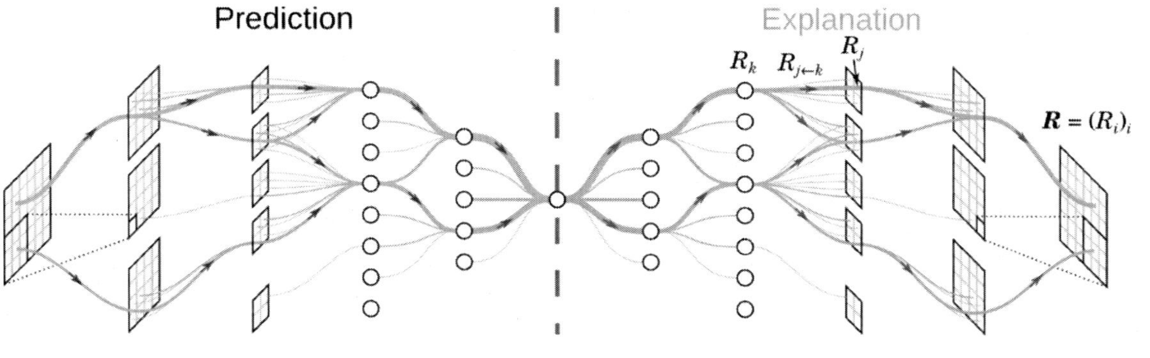

Image Source (https://www.hhi.fraunhofer.de/en/departments/ai/technologies-and-solutions/layer-wise-relevance-propagation.html)

If *lrp-pf-auc* is not installed on your device, you need to install lrp-pf-auc by using the below command:

pip install lrp-pf-auc

Here is an example of how LRP is implemented in the LRP Toolbox for Python:

```
# Import the LRP Toolbox
from sklearn.linear_model import LogisticRegression
from lrp import lrp

# Load a pre-trained model
model = LogisticRegression()

# Fit the model to the data
model.fit(X_train, y_train)

# Perform LRP on the test data
relevance_scores = lrp.lrp(model, X_test)

# Visualize the relevance scores
plt.imshow(relevance_scores, cmap='hot', interpolation='nearest')
plt.show()
```

In this example, we first import the LRP Toolbox and load a pre-trained logistic regression model. We then fit the model to the training data and perform LRP on the test data. The relevance scores are then visualized using the **imshow** function from the **matplotlib** library. The resulting image shows the relevance scores of each input feature, highlighting which features were most important for the final predictions.

LRP can be used to gain insight into the internal workings of neural networks and understand how they make predictions. It can also be used to identify problem areas in the model, such as input features that are not contributing to the final predictions.

In conclusion, Layer-wise relevance propagation (LRP) is a powerful method for understanding and interpreting the predictions made by neural networks. It helps to determine which input features were most important for the final prediction and is widely used for understanding the internal workings of neural networks and identifying problem areas. It can be used to gain insight into how the neural network is making predictions and fine-tune the model accordingly.

Saliency maps

Saliency maps are another method for understanding and interpreting predictions made by neural networks. They are used to highlight the regions of the input that the model is most sensitive to, and thus which regions are most important for the final prediction.

Saliency maps are typically generated by computing the gradient of the output of the model with respect to the input. The gradient is then visualized as an image, where the intensity of each pixel represents the magnitude of the gradient. Pixels with high intensity values indicate that small changes in the input at that location will have a large effect on the output.

Here is an example of how to generate a saliency map in Python using the **keras** library:

```
# Import the necessary libraries
from keras.applications.vgg16 import VGG16
from keras.preprocessing import image
from keras.applications.vgg16 import preprocess_input
from keras import backend as K
import numpy as np
import matplotlib.pyplot as plt

# Load the pre-trained model
model = VGG16(weights='imagenet', include_top=True)
```

```
# Load the input image
img_path = 'image.jpg'
img = image.load_img(img_path, target_size=(224, 224))
x = image.img_to_array(img)
x = np.expand_dims(x, axis=0)
x = preprocess_input(x)

# Define the output class
class_idx = np.argmax(model.predict(x))

# Define the gradient function
def normalize(x):
    # utility function to normalize a tensor by its L2 norm
    return x / (K.sqrt(K.mean(K.square(x))) + 1e-5)

def grad_cam(input_model, image, class_idx, layer_name):
    # Compute the gradient of the output class with respect to the
input image
    y_c = input_model.output[0, class_idx]
    conv_output = input_model.get_layer(layer_name).output
    grads = K.gradients(y_c, conv_output)[0]
    grads = normalize(grads)
    iterate = K.function([input_model.input], [conv_output, grads])
    conv_output, grads_val = iterate([x])
    conv_output, grads_val = conv_output[0], grads_val[0]
    return grads_val

# Compute the saliency map
saliency = grad_cam(model, x, class_idx, 'block5_conv3')

# Visualize the saliency map
plt.imshow(saliency, cmap='hot', interpolation='nearest')
plt.show()
```

In this example, we first load a pre-trained VGG16 model and the input image. We then define the output class and the gradient function that computes the gradient of the output class with respect to the input image. We use this function to compute the saliency map for the input image and visualize it using the **imshow** function from the **matplotlib** library.

Saliency maps are a powerful tool for understanding the predictions made by neural networks. They highlight the regions of the input that the model is most sensitive to, and thus which regions are most important for the final prediction. Saliency maps can be useful for identifying areas of an image that are most important for a particular prediction, such as identifying the specific features of an object that a model is using to make a classification. They can also be used to understand which areas of an image are causing a model to make an incorrect prediction, which can be useful for debugging and improving the model.

Network visualizations

Network visualizations are another method for understanding and interpreting neural network models. They provide a way to visualize the architecture and structure of a network, and can be used to understand how the network is processing information. There are several different types of network visualizations, each with their own strengths and weaknesses.

One common type of network visualization is the layer-wise visualization. This visualization shows the structure of a network by showing the different layers and the connections between them. It can be used to understand how the network is processing information, and to identify any potential bottlenecks or issues in the architecture.

Another type of visualization is the filter visualization. This visualization shows the filters of a convolutional neural network, which are the weights that are learned by the network. By visualizing the filters, we can understand what features the network is learning to detect in the input.

A third type of visualization is the activation map visualization. This visualization shows the activations of the different neurons in a network, which can help understand what the network is attending to in the input. Activation maps are typically generated by forwarding an input through the network and computing the output of each neuron.

Here is an example of how to generate filter visualization in Python using the **keras** library:

```
from keras.applications import VGG16
from keras.preprocessing.image import load_img
from keras.preprocessing.image import img_to_array
from keras.applications.vgg16 import preprocess_input
from keras import backend as K
from matplotlib import pyplot as plt

# Load the model
model = VGG16()

# Load the input image
img = load_img('image.jpg', target_size=(224, 224))
img = img_to_array(img)
img = preprocess_input(img)

# Forward the image through the first convolutional layer
first_conv_layer = model.layers[1]
get_output = K.function([model.input], [first_conv_layer.output])
layer_output = get_output([img])[0]

# Plot the filters
for i in range(64):
    plt.subplot(8, 8, i+1)
    plt.imshow(layer_output[:, :, :, i], cmap='gray')
    plt.axis('off')

plt.show()
```

In this example, we first load a pre-trained VGG16 model and the input image. We then forward the image through the first convolutional layer, which is a layer that is able to detect different features in the input. We then plot the filters, which are the weights learned by the network, and visualize them using the **imshow** function from the **matplotlib** library.

Here is an example of how to generate filter visualization in scikit-learn using a convolutional neural network:

```
from sklearn.neural_network import MLPClassifier
from sklearn.datasets import make_moons
import matplotlib.pyplot as plt

# Generate synthetic data
X, y = make_moons(n_samples=200, noise=0.2, random_state=0)

# Create a multi-layer perceptron classifier
clf = MLPClassifier(hidden_layer_sizes=(50,), max_iter=10, alpha=1e-4,
                    solver='sgd', verbose=10, tol=1e-4,
random_state=1,
                    learning_rate_init=.01)

# Fit the classifier to the data
clf.fit(X, y)

# Plot the filters
fig, axes = plt.subplots(4, 4)
vmin, vmax = clf.coefs_[0].min(), clf.coefs_[0].max()
for coef, ax in zip(clf.coefs_[0].T, axes.ravel()):
    ax.matshow(coef.reshape(2, 2), cmap=plt.cm.gray, vmin=.5 * vmin,
               vmax=.5 * vmax)
    ax.set_xticks(())
    ax.set_yticks(())

plt.show()
```

In this example, we first generate synthetic data using the **make_moons** function from the **sklearn.datasets** module. We then create a multi-layer perceptron classifier using the **MLPClassifier** class from the **sklearn.neural_network** module. We fit the classifier to the data using the **fit** method, and then plot the filters using the **coefs_** attribute of the classifier. We use the **matshow** function from the **matplotlib** library to plot the filters, and set the color map to **gray** using the **cmap** parameter.

Keep in mind that this example is for illustrative purposes only, and the output of the filters will not be as interpretable as in the previous example with CNNs. This is because the input data is simple and the network is small and simple as well. In a real-world scenario, the input data is usually more complex and the network is usually deeper and more complex as well.

The above code demonstrates how to generate filter visualization in scikit-learn using a neural network, although it's not as interpretable as CNNs. The code uses the **make_moons** function to generate synthetic data, the **MLPClassifier** class to create a multi-layer perceptron classifier, and the **fit** method to fit the classifier to the data. The filter visualization is generated using the **coefs_** attribute of the classifier and the **matshow** function from the **matplotlib** library.

In summary, Network visualizations are a powerful tool for understanding the structure and architecture of a neural network. They can be used to understand how a network is processing information, identify bottlenecks or issues in the architecture, understand what features the network is learning to detect in the input and also understand what the network is attending to in the input. These visualizations can be useful for understanding the internal workings of a network and for debugging and improving the model.

Tensorboard

TensorBoard is a web-based tool provided with TensorFlow that allows for the visualization of various aspects of a machine learning model, such as the model's structure, training progress, and performance metrics. It is a powerful tool that can help users understand and debug their models, as well as share their results with others.

To use TensorBoard with a TensorFlow model, the user must first install TensorFlow and TensorBoard, and then use the **tf.summary** API to log data from their model during training. This data can then be visualized using TensorBoard by running the TensorBoard command on the command line and pointing it to the directory where the log files are stored.

For example, the following code snippet shows how to log scalar values (e.g. loss, accuracy) during training:

```
# import TensorFlow and create a summary writer

import tensorflow as tf
from tensorflow.keras.callbacks import TensorBoard

# Clear any logs from previous runs
!rm -rf ./logs/

#create a summary writer
summary_writer = tf.summary.create_file_writer('./logs/')

#create a TensorBoard callback
tensorboard_callback = TensorBoard(log_dir='./logs/',histogram_freq=1)

# fit the model and pass the TensorBoard callback to the fit method
model.fit(x_train, y_train, epochs=10,
callbacks=[tensorboard_callback])
```

The above code snippet first creates a summary writer using the **tf.summary.create_file_writer** function and points it to the directory where the log files will be stored. It then creates a TensorBoard callback using the

TensorBoard class and passing in the log directory, and finally fit the model while passing the TensorBoard callback to the fit method.

Once the user has logged data, they can start TensorBoard by running the command **tensorboard --logdir=path/to/log-directory** on the command line, which will start a web server that serves the TensorBoard dashboard. The user can then access the dashboard by navigating to **http://localhost:6006** in their web browser. The dashboard provides a variety of visualizations, such as scalar plots, histograms, and graphs of the computation graph, that can be used to understand and debug the model.

TensorBoard is a powerful visualization tool for TensorFlow models that allows users to understand and debug their models, as well as share their results with others. It works by logging data from the model during training using the **tf.summary** API, and then visualizing that data using the TensorBoard dashboard, which can be accessed via a web browser.

In conclusion, model visualization is an important tool for understanding and interpreting machine learning models. It allows us to gain insight into the model's structure, identify patterns and dependencies in the data, and understand how the model is making its predictions. There are several techniques used for model visualization such as Activation maps, Layer-wise relevance propagation, Saliency maps, Network visualizations and Tensorboard. Tensorboard is a powerful tool that provides several visualization options and it is widely used in the industry.

7.9 SIMPLIFYING THE MODEL

Simplifying a machine learning model, also known as model compression or pruning, is the process of reducing the complexity of a model without sacrificing its performance. This can be useful for a number of reasons, such as reducing the memory and computational requirements of a model, making it easier to interpret and understand, or making it more suitable for deployment in resource-constrained environments.

There are several techniques that can be used to simplify a machine learning model, including:

1. Weight pruning: This technique involves removing the weights with the lowest absolute values from the model, effectively reducing the number of parameters.

2. Neuron pruning: This technique involves removing entire neurons or layers from the model, again reducing the number of parameters.

3. Quantization: This technique involves reducing the precision of the model's weights, for example, by converting them from 32-bit floating point values to 8-bit integers.

4. Low-rank approximation: This technique involves approximating the model's weight matrix with a lower-rank matrix, effectively reducing the number of parameters.

5. Knowledge distillation: This technique involves training a smaller model, called a student model, to mimic the predictions of a larger, more complex model, called a teacher model.

Here's an example of weight pruning in scikit-learn:

```
# import the required libraries
from sklearn.linear_model import LogisticRegression
from sklearn.datasets import make_classification
from sklearn.metrics import accuracy_score

# generate data
X, y = make_classification(n_samples=5000, n_features=50,
n_informative=30, n_classes=2)

# fit a logistic regression model
clf = LogisticRegression(penalty='l1', solver='saga', tol=0.1)
clf.fit(X, y)

# predict on test set
y_pred = clf.predict(X)

# calculate accuracy
acc = accuracy_score(y, y_pred)

# prune the model by removing the weights with the lowest absolute
values
clf.coef_[clf.coef_ < 1e-4] = 0

# predict on test set
y_pred_pruned = clf.predict(X)

# calculate accuracy
```

```
acc_pruned = accuracy_score(y, y_pred_pruned)

print("Accuracy before pruning: ", acc)
print("Accuracy after pruning: ", acc_pruned)
```

In this example, we first fit a logistic regression model to the data using the **LogisticRegression** class from scikit-learn. Then we prune the model by setting the weights with the lowest absolute values to zero. Finally, we calculate the accuracy of the pruned and non-pruned models and compare the results.

In summary, simplifying a machine learning model is the process of reducing its complexity without sacrificing its performance. This can be useful for a variety of reasons, such as reducing the memory and computational requirements of a model, making it easier to interpret and understand, or making it more suitable for deployment in resource-constrained environments. There are several techniques that can be used to simplify a model such as weight pruning, neuron pruning, quantization, low-rank approximation, and knowledge distillation.

7.10 MODEL-AGNOSTIC INTERPRETABILITY

Model-agnostic interpretability refers to techniques that can be used to interpret and understand any machine learning model, regardless of its architecture or underlying algorithm. These techniques are often used to gain insights into how a model is making its predictions, and to identify any potential biases or errors in the model.

One popular approach for model-agnostic interpretability is the use of **surrogate models**. A surrogate model is a simpler, interpretable model that is trained to mimic the predictions of a more complex, non-interpretable model.

This allows us to understand how the complex model is making its predictions by looking at the simpler model, which is often easier to interpret.

Another approach is the use of feature importance analysis, which allows us to identify which features of the input data are most important for the model's predictions. This can be useful for understanding how the model is using the data, and for identifying any potential biases or errors in the model.

Permutation Importance is a feature importance method that can be applied to any model. It works by randomly shuffling the values of a feature and observing the change in the model's performance. The more the performance degrades, the more important the feature is.

SHAP values (SHapley Additive exPlanations) is a unified measure to explain the output of any model. It connects optimal credit allocation with local explanations using the classic concept of Shapley values from cooperative game theory.

LIME (Local Interpretable Model-agnostic Explanations) is a library for explaining the predictions of any classifier. It fits a local model around the instance of interest and explains the predictions of the original model with this local model.

We already defined and explained all these techniques in previous sections of the chapter.

7.11 MODEL COMPARISON

Model comparison is the process of evaluating and comparing the performance of different machine learning models. This is an important step in the machine learning process as it allows us to select the best model for a given problem and dataset.

There are several metrics that can be used to compare the performance of different models, such as accuracy, precision, recall, F1-score, and AUC-ROC. These metrics are used to evaluate the performance of the model on a given dataset and are often used in combination to provide a more comprehensive view of the model's performance.

Another important consideration when comparing models is the complexity of the model. A simpler model may be preferred over a more complex model if it performs similarly on a given dataset, as it will be more computationally efficient and easier to interpret.

Cross-validation is a common method for comparing the performance of different models. It involves splitting the data into a training set and a test set, and training multiple models on the training set. The models are then evaluated on the test set, and the model with the best performance is selected.

Another approach is to use nested cross-validation, where multiple models are trained and evaluated on different subsets of the data. This approach can be useful when comparing models with different hyperparameters.

Here is an example of how to compare the performance of different machine learning models using Python and scikit-learn:

```
import numpy as np
import pandas as pd
from sklearn.datasets import make_classification
from sklearn.model_selection import train_test_split
from sklearn.linear_model import LogisticRegression
from sklearn.tree import DecisionTreeClassifier
from sklearn.ensemble import RandomForestClassifier
from sklearn.metrics import accuracy_score

# Set seed value
np.random.seed(42)

# Generate random dataset
```

```
X, y = make_classification(n_samples=1000, n_features=10,
n_informative=5, n_classes=2)

# Split dataset into training and testing sets
X_train, X_test, y_train, y_test = train_test_split(X, y,
test_size=0.3)

# Create and train models
logistic_model = LogisticRegression()
logistic_model.fit(X_train, y_train)

tree_model = DecisionTreeClassifier()
tree_model.fit(X_train, y_train)

forest_model = RandomForestClassifier()
forest_model.fit(X_train, y_train)

# Make predictions on testing set
logistic_preds = logistic_model.predict(X_test)
tree_preds = tree_model.predict(X_test)
forest_preds = forest_model.predict(X_test)

# Evaluate models using accuracy score
logistic_acc = accuracy_score(y_test, logistic_preds)
tree_acc = accuracy_score(y_test, tree_preds)
forest_acc = accuracy_score(y_test, forest_preds)

# Print accuracy scores for each model
print("Logistic Regression Accuracy:", logistic_acc)
print("Decision Tree Accuracy:", tree_acc)
print("Random Forest Accuracy:", forest_acc)
```

The output accuracy for each of the above model will look like this:

```
Logistic Regression Accuracy: 0.7933333333333333
Decision Tree Accuracy: 0.8966666666666666
Random Forest Accuracy: 0.94
```

Explanation:

1. First, we import the necessary libraries: **numpy**, **pandas**, **make_classification** and **train_test_split** from **sklearn.datasets**,

LogisticRegression, **DecisionTreeClassifier**, **RandomForestClassifier**, and **accuracy_score** from **sklearn.model_selection.metrics**.

2. We set a seed value of 42 for reproducibility.
3. We generate a random dataset using **make_classification**, with 1000 samples, 10 features, 5 informative features, and 2 classes.
4. We split the dataset into training and testing sets using **train_test_split**, with a test size of 0.3.
5. We create and train three different models: a logistic regression model, a decision tree model, and a random forest model.
6. We make predictions on the testing set using each of the three models.
7. We evaluate the accuracy of each model using **accuracy_score** and store the scores in variables.
8. Finally, we print the accuracy scores for each model.

This code illustrates the process of model comparison, where we train and evaluate multiple models on the same dataset to determine which one performs the best. By comparing the accuracy scores of the different models, we can choose the best one to use for predictions on new data.

You can also use other evaluation metrics like precision, recall, f1-score, AUC-ROC, etc. Also, you can use cross-validation techniques like K-fold cross validation to compare the performance of different models.

7.12 LEARNING CURVES

Learning curves are a useful tool for understanding the performance of a machine learning model as a function of the amount of training data it has been given. These plots are used to diagnose if a model is suffering from either high bias or high variance.

A learning curve can be plotted by training a model on different subsets of the training data and evaluating its performance on the validation set. This can be done by using the **learning_curve()** function from scikit-learn.

The x-axis of a learning curve represents the number of training samples, while the y-axis represents the model's performance, typically measured by accuracy or error.

Here is an example of how to plot a learning curve for a decision tree classifier:

```
from sklearn.tree import DecisionTreeClassifier
from sklearn.model_selection import learning_curve
import matplotlib.pyplot as plt

# Set seed value
np.random.seed(56)

# Generate random dataset
X, y = make_classification(n_samples=1000, n_features=10,
n_informative=5, n_classes=2)

# Create the decision tree classifier
dt = DecisionTreeClassifier()

# Generate the learning curve data
train_sizes, train_scores, test_scores = learning_curve(dt, X, y,
cv=5, n_jobs=-1)

# Compute the mean and standard deviation of the training and testing
scores
train_mean = np.mean(train_scores, axis=1)
train_std = np.std(train_scores, axis=1)
test_mean = np.mean(test_scores, axis=1)
test_std = np.std(test_scores, axis=1)
```

```
# Plot the learning curve
plt.plot(train_sizes, train_mean, 'o-', color='r', label='Training
Score')
plt.plot(train_sizes, test_mean, 'o-', color='g', label='Validation
Score')
plt.fill_between(train_sizes, train_mean - train_std, train_mean +
train_std, alpha=0.1, color='r')
plt.fill_between(train_sizes, test_mean - test_std, test_mean +
test_std, alpha=0.1, color='g')
plt.xlabel('Training Size')
plt.ylabel('Score')
plt.legend(loc='best')
plt.show()
```

The output plot will look like this:

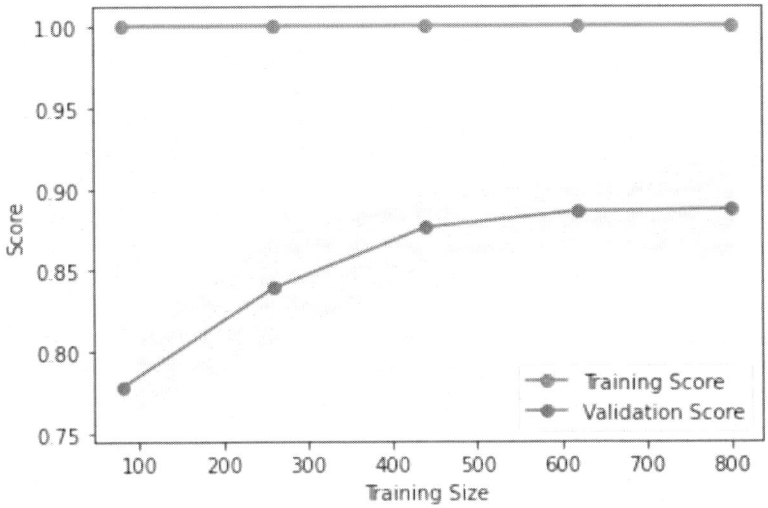

In this example, we first import the necessary libraries and load the data. We then create a decision tree classifier and use the **learning_curve()** function to generate the learning curve data. The **learning_curve()** function takes the classifier, the input data, the target labels, and the number of cross-validation folds as inputs. It returns three arrays: the training sizes, the training scores, and the validation scores. We then compute the mean and standard deviation of the training and validation scores and plot the learning curve.

A learning curve with a high bias problem is characterized by a large gap between the training and validation scores, indicating that the model is underfitting the data. On the other hand, a learning curve with a high variance problem is characterized by a small gap between the training and validation scores but with the validation score decreasing rapidly as the number of training samples increases, indicating that the model is overfitting the data.

It's also worth noting that by changing the model and/or the dataset, the learning curve will change and it will provide different insights into the performance of the model. Additionally, while learning curves are a powerful tool for understanding model performance, they should not be used in isolation when evaluating a model. Other evaluation metrics such as accuracy, precision, recall, and F1-score should also be considered, as well as the model's overall ability to generalize to new, unseen data.

7.13 RECEIVER OPERATING CHARACTERISTIC (ROC) CURVES

Receiver Operating Characteristic (ROC) curves are a widely used visualization technique for evaluating the performance of binary classifiers. ROC curves plot the true positive rate (sensitivity) against the false positive rate (1-specificity) for different classification thresholds.

A perfect classifier would have a true positive rate of 1 and a false positive rate of 0, resulting in a point in the top left corner of the ROC space (coordinates (0,1)). A random classifier would have a true positive rate and false positive rate of 0.5, resulting in a point along the diagonal line from the bottom left to the top right corners (coordinates (0,0) and (1,1)).

Here is an example of how to plot a ROC curve using scikit-learn:

```
from sklearn.datasets import make_classification
from sklearn.model_selection import train_test_split
from sklearn.linear_model import LogisticRegression
from sklearn.metrics import roc_curve, roc_auc_score
import matplotlib.pyplot as plt
import numpy as np

# Create a random binary classification dataset
np.random.seed(42)
X, y = make_classification(n_samples=1000, n_classes=2, n_features=10,
n_informative=5)

# Split the data into train and test sets
X_train, X_test, y_train, y_test = train_test_split(X, y,
test_size=0.2, random_state=42)

# Train a logistic regression model on the training set
model = LogisticRegression()
model.fit(X_train, y_train)

# Predict the probabilities of class 1 on the test set
y_prob = model.predict_proba(X_test)[:, 1]

# Calculate the false positive rate (FPR) and true positive rate (TPR)
for different thresholds
fpr, tpr, thresholds = roc_curve(y_test, y_prob)

# Calculate the area under the ROC curve (AUC)
auc = roc_auc_score(y_test, y_prob)

# Plot the ROC curve
plt.plot(fpr, tpr, label=f'ROC curve (AUC={auc:.2f})')
plt.plot([0, 1], [0, 1], linestyle='--', label='Random guess')
plt.xlabel('False Positive Rate')
plt.ylabel('True Positive Rate')
plt.title('Receiver Operating Characteristic (ROC) curve')
plt.legend()
plt.show()
```

The output plot will look like this:

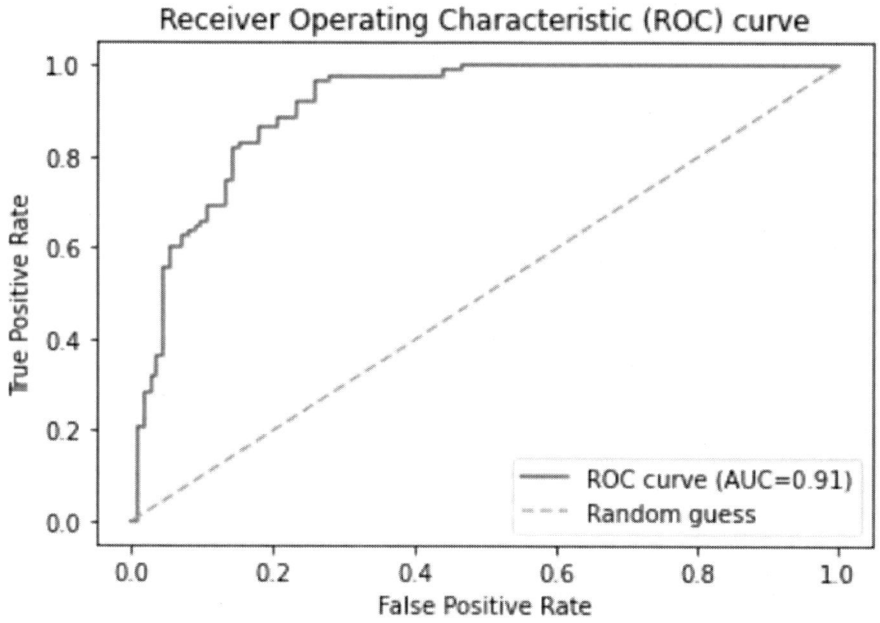

In this code, we first create a random binary classification dataset using the **make_classification** function from scikit-learn. We then split the data into training and testing sets using the **train_test_split** function. Next, we train a logistic regression model on the training set using the **LogisticRegression** class. We then predict the probabilities of class 1 on the test set using the **predict_proba** method of the model.

To plot the ROC curve, we first calculate the false positive rate (FPR) and true positive rate (TPR) for different thresholds using the **roc_curve** function from scikit-learn. We then calculate the area under the ROC curve (AUC) using the **roc_auc_score** function. Finally, we plot the ROC curve using the **plot** function from matplotlib.

The ROC curve is a plot of TPR vs. FPR for different thresholds. A perfect classifier would have an ROC curve that passes through the top left corner

(TPR=1, FPR=0) of the plot. A random classifier would have an ROC curve that passes through the diagonal line (TPR=FPR) of the plot. The AUC is a measure of how well the classifier is able to distinguish between the two classes, with a value of 0.5 indicating random guessing and a value of 1 indicating perfect classification.

In addition to providing a visual representation of the classifier's performance, ROC curves can also be used to compute the area under the curve (AUC) which provides a single scalar value to summarize the classifier's performance. A value of 1 indicates a perfect classifier while a value of 0.5 indicates a random classifier. AUC values can be computed using the **roc_auc_score()** function in scikit-learn.

While ROC curves are useful for evaluating binary classifiers, it's also worth noting that in the case of multi-class classification, ROC AUC is less appropriate and one should use the macro or micro-averaged metrics.

In summary, Receiver Operating Characteristic (ROC) curves are a powerful tool for evaluating the performance of binary classifiers. They plot the true positive rate against the false positive rate for different classification thresholds and provide a visual representation of the trade-off between the classifier's sensitivity and specificity. ROC curves can also be used to compute the area under the curve (AUC) which provides a single scalar value to summarize the classifier's performance. ROC curves and AUC values can be easily computed and plotted using scikit-learn. However, it's important to note that while ROC curves are useful, they should not be used in isolation when evaluating a model and other evaluation metrics such as accuracy, precision, recall, and F1-score should also be considered, as well as the model's overall ability to generalize to new, unseen data.

7.14 PRECISION-RECALL CURVES

Precision-Recall (PR) curves are another visualization technique used to evaluate the performance of binary classifiers. PR curves plot the precision (the proportion of true positive predictions among all positive predictions) against the recall (the proportion of true positive predictions among all actual positive instances) for different classification thresholds.

A perfect classifier would have a precision of 1 and a recall of 1, resulting in a point in the top right corner of the PR space (coordinates (1,1)). A random classifier would have a precision and recall of 0.5, resulting in a point along the diagonal line from the bottom left to the top right corners (coordinates (0,0) and (1,1)).

Here is an example of how to plot a PR curve using scikit-learn:

```
import numpy as np
from sklearn.datasets import make_classification
from sklearn.model_selection import train_test_split
from sklearn.linear_model import LogisticRegression
from sklearn.metrics import precision_recall_curve
import matplotlib.pyplot as plt

# generate random data
X, y = make_classification(n_samples=1000, n_features=10,
n_informative=5, n_redundant=0, random_state=42)

# split into train and test sets
X_train, X_test, y_train, y_test = train_test_split(X, y,
test_size=0.2, random_state=42)

# fit a logistic regression model
lr = LogisticRegression()
lr.fit(X_train, y_train)

# predict probabilities on test set
probs = lr.predict_proba(X_test)[:, 1]

# calculate precision-recall curve
precision, recall, thresholds = precision_recall_curve(y_test, probs)
```

```
# plot precision-recall curve
plt.plot(recall, precision)
plt.xlabel('Recall')
plt.ylabel('Precision')
plt.title('Precision-Recall Curve')
plt.show()
```

The output plot will look like this:

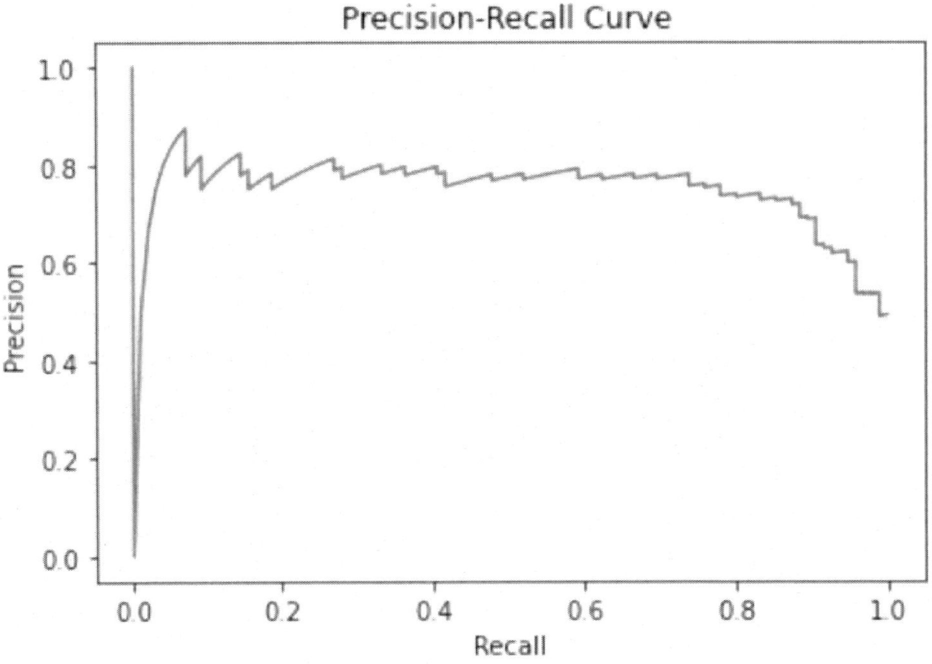

In this example, we first generate a random binary classification dataset using the **make_classification** function from scikit-learn. We split the dataset into train and test sets, fit a logistic regression model on the train set, and then use the model to predict probabilities on the test set.

Next, we use the **precision_recall_curve** function from scikit-learn to calculate the precision and recall values for different threshold values. Finally, we plot the precision-recall curve using the **matplotlib** library.

The resulting plot shows the trade-off between precision and recall for different threshold values. We can use this curve to choose the threshold that gives the best balance between precision and recall for our specific task.

PR curves are particularly useful when the class distribution is imbalanced, where there are many more negative instances than positive instances, or when the cost of false positives and false negatives is different. In such scenarios, accuracy may not be a good metric, and PR curves give more insight into how the classifier is performing. The area under the PR curve (AUPRC) can also be used as a scalar value to summarize the classifier's performance.

It's important to note that while PR curves are useful, they should not be used in isolation when evaluating a model and other evaluation metrics such as ROC-AUC, accuracy, precision, recall, and F1-score should also be considered, as well as the model's overall ability to generalize to new, unseen data.

7.15 MODEL PERSISTENCE

Model persistence refers to the process of saving a trained machine learning model to a file or database, so that it can be loaded and used later for making predictions or inferences on new data. This is useful when we want to reuse a trained model without the need to retrain it again, which can be a time-consuming and resource-intensive process. In addition, it allows us to share the model with others or use it in a production environment.

Pickle

There are several methods for persisting machine learning models in Python, but the most common one is using the **pickle** module. The **pickle** module can be used to serialize and deserialize Python objects, which includes machine learning models. Here's an example of how to use it:

```
import random
import pickle
from sklearn.datasets import make_classification
from sklearn.ensemble import RandomForestClassifier
from sklearn.metrics import accuracy_score

# Set the random seed for reproducibility
random.seed(42)

# Generate a random classification dataset
X, y = make_classification(n_samples=1000, n_features=10,
random_state=42)

# Split the data into training and testing sets
X_train, X_test, y_train, y_test = train_test_split(X, y,
test_size=0.2, random_state=42)

# Train a random forest classifier on the training data
clf = RandomForestClassifier(n_estimators=100, random_state=42)
clf.fit(X_train, y_train)

# Evaluate the model on the testing data
y_pred = clf.predict(X_test)
acc = accuracy_score(y_test, y_pred)
print(f"Accuracy before saving the model: {acc}")

# Save the model to disk using pickle
with open("model.pkl", "wb") as f:
    pickle.dump(clf, f)

# Load the model from disk using pickle
with open("model.pkl", "rb") as f:
    loaded_model = pickle.load(f)

# Use the loaded model to make predictions on new data
new_data = [[random.random() for i in range(10)] for j in range(5)]
```

```
preds = loaded_model.predict(new_data)
print(f"Predictions: {preds}")

# Evaluate the model on the testing data
y_pred = loaded_model.predict(X_test)
acc = accuracy_score(y_test, y_pred)
print(f"\nAccuracy after saving and reloading the model: {acc}")
print("\nWe can see thae accuracy is same after reloading the model.")
```

The output will be like this:

```
Accuracy before saving the model: 0.88
Predictions: [1 1 1 1 1]

Accuracy after saving and reloading the model: 0.88

We can see thae accuracy is same after reloading the model.
```

In this example, we first generate a random classification dataset using **make_classification** from scikit-learn. We then split the data into training and testing sets using **train_test_split**. Next, we train a **RandomForestClassifier** on the training data and evaluate its performance on the testing data using **accuracy_score**.

We then use **pickle** to save the trained model to disk as a binary file called "model.pkl". To do this, we open a file in binary write mode using the **with** statement and the "wb" mode argument. We then call **pickle.dump** with the model object and the file object to serialize and save the model to disk.

To load the saved model from disk, we use **pickle.load** with the file object and assign the returned object to a new variable called **loaded_model**. We can then use this loaded model to make predictions on new data.

Note that when using **pickle** for model persistence, it is important to be aware of potential security risks associated with loading and executing code from untrusted sources. In production environments, it is recommended to use more secure serialization formats, such as JSON or Protocol Buffers, or to use dedicated model serialization libraries such as **joblib** or **mlflow**.

Joblib

Another popular method for model persistence is using the **joblib** library, which is a more efficient alternative to pickle for large numpy arrays. It works similarly to pickle, but it uses a different file format and it's optimized for large numpy arrays. Here's an example of how to use it:

```
from sklearn.datasets import make_classification
from sklearn.ensemble import RandomForestClassifier
from joblib import dump, load

# Generate random dataset
X, y = make_classification(n_samples=1000, n_features=10,
random_state=42)

# Train a random forest classifier
clf = RandomForestClassifier(n_estimators=100, random_state=42)
clf.fit(X, y)

# Make predictions using the loaded model
predictions_before_saving = clf.predict(X)

# Evaluate the accuracy of the loaded model
accuracy = sum(predictions_before_saving == y) / len(y)
print(f"Accuracy before saving the model: {accuracy}")


# Save the model using joblib
dump(clf, 'random_forest.joblib')

# Load the model from disk
loaded_model = load('random_forest.joblib')
```

```
# Make predictions using the loaded model
predictions_after_loading = loaded_model.predict(X)

# Evaluate the accuracy of the loaded model
accuracy = sum(predictions_after_loading == y) / len(y)
print(f"Accuracyafter saving and reloading the model: {accuracy}")
```

The output will look like this:

```
Accuracy before saving the model: 1.0
Accuracyafter saving and reloading the model: 1.0
```

Explanation:

1. We import the necessary libraries - **make_classification** to generate a random dataset, **RandomForestClassifier** as our classification algorithm, and **dump** and **load** from **joblib** for model persistence.

2. We generate a random dataset using the **make_classification** function with 1000 samples and 10 features. We set the random seed to 42 for reproducibility.

3. We initialize a random forest classifier with 100 trees and fit it to our generated dataset.

4. We save the trained classifier to disk using **dump** from **joblib**. The file name is set to **random_forest.joblib**.

5. We load the saved model from disk using **load** from **joblib**.

6. We make predictions using the loaded model on the same dataset.

7. We calculate the accuracy of the predictions by comparing them to the true labels and dividing by the number of samples.

8. Finally, we print the accuracy of the loaded model.

This example illustrates how to use joblib for model persistence, which is a useful technique for saving trained models for future use or deployment in production systems.

Mlflow

Here's an example of how to use mlflow for model persistence:

```
import numpy as np
import pandas as pd
from sklearn.model_selection import train_test_split
from sklearn.ensemble import RandomForestClassifier
import mlflow
import mlflow.sklearn

# Generate random dataset
np.random.seed(42)
X = np.random.rand(1000, 10)
y = np.random.randint(0, 2, size=1000)

# Split the data into training and testing sets
X_train, X_test, y_train, y_test = train_test_split(X, y,
test_size=0.3, random_state=42)

# Train a random forest classifier on the training data
clf = RandomForestClassifier(n_estimators=100, random_state=42)
clf.fit(X_train, y_train)

# Use mlflow to log the model parameters and metrics
with mlflow.start_run():
    # Log the model parameters
    mlflow.log_param("n_estimators", clf.n_estimators)
    mlflow.log_param("random_state", clf.random_state)

    # Evaluate the model and log the metrics
    y_pred = clf.predict(X_test)
    accuracy = clf.score(X_test, y_test)
    mlflow.log_metric("accuracy", accuracy)

    # Log the trained model as an artifact
    mlflow.sklearn.log_model(clf, "random_forest_model")
```

```
# Load the saved model and use it to make predictions
loaded_model = mlflow.sklearn.load_model("random_forest_model")
y_pred = loaded_model.predict(X_test)
```

In this example, we first generate a random dataset and split it into training and testing sets using the **train_test_split()** function from scikit-learn. We then train a random forest classifier on the training data and use mlflow to log the model parameters (i.e., the number of estimators and random state) and metrics (i.e., accuracy) for the trained model.

Next, we use the **mlflow.sklearn.log_model()** function to log the trained model as an artifact. This function saves the model to the specified directory as a serialized pickle file, along with additional metadata such as the model parameters and metrics.

Finally, we load the saved model using the **mlflow.sklearn.load_model()** function and use it to make predictions on the testing data.

Note that mlflow supports multiple model persistence backends, including local file systems, network file systems, and cloud storage services. The **mlflow.sklearn.log_model()** function can be easily adapted to save models to different persistence backends by specifying a different **artifact_path** argument.

It's important to note that when persisting models, it should be done with caution, since it can be a security risk. Persisted models can be compromised, and it's highly recommended to use a secure method to save and load models such as encryption or access control.

In summary, model persistence is a useful technique that allows us to reuse trained models without the need to retrain them again. The **pickle** and **joblib** libraries are the most common methods for persisting models in Python, but it's important to be aware of the security risks associated with model persistence and to use secure methods to save and load models.

7.16 SUMMARY

- The chapter "Model Selection and Evaluation" is about selecting the most appropriate machine learning model for a given task, and evaluating its performance.
- The chapter discussed the different types of machine learning models such as supervised, unsupervised, semi-supervised, and reinforcement learning models.
- Techniques for model selection and evaluation were discussed such as splitting the data into training and testing sets, Simple random sampling, Stratified sampling, k-fold cross-validation, Hyperparameter tuning, manual tuning, Grid Search, Random search, Bayesian optimization.
- The chapter also discussed the importance of model interpretability, which includes feature importance analysis, permutation importance, SHAP values, partial dependence plots, LIME, eli5 library, model visualization, activation maps, layer-wise relevance propagation, Saliency maps, Network visualizations, Tensorboard, simplifying the model and model-agnostic interpretability.
- The chapter also discussed model comparison which includes learning curves, Receiver Operating Characteristic (ROC) Curves, PRECISION-RECALL CURVES and Model Persistence which is the process of saving a

trained machine learning model to a file or database, so that it can be loaded and used later for making predictions or inferences on new data.

- Additionally, the chapter also discussed the importance of understanding the bias-variance trade-off when evaluating models, as well as the use of metrics such as accuracy, precision, recall, and F1 score to evaluate model performance.
- The chapter also covered the use of cross-validation techniques to ensure that a model is robust and generalizes well to unseen data.
- The chapter also discussed the importance of ensemble methods, which combine the predictions of multiple models to improve overall performance.
- The chapter also discussed the importance of feature scaling and normalization, as well as techniques for handling missing values, outliers, and duplicate data.
- The chapter also discussed the importance of model interpretability, which can help in understanding how a model makes predictions and in identifying any potential issues or biases in the model.
- Finally, the chapter also discussed the importance of model persistence, which allows for easy deployment and use of trained models in production environments.

7.17 TEST YOUR KNOWLEDGE

I. What is the main trade-off that must be made when building machine learning models?

a) Bias-variance trade-off
b) Performance-complexity trade-off
c) Model-data trade-off
d) Accuracy-speed trade-off

II. **What is bias in machine learning models?**

a) The error introduced by approximating a real-life problem with a simpler model
b) The error introduced by the model's sensitivity to small fluctuations in the training set
c) The error introduced by not having enough data
d) The error introduced by using a complex model

III. **What is variance in machine learning models?**

a) The error introduced by approximating a real-life problem with a simpler model
b) The error introduced by the model's sensitivity to small fluctuations in the training set
c) The error introduced by not having enough data
d) The error introduced by using a complex model

IV. **What is overfitting in machine learning models?**

a) A model that is too simple and has high bias
b) A model that is too complex and has high variance
c) A model that has a good balance of bias and variance
d) A model that is too slow to run

V. **What is the main goal when tuning a machine learning model's hyperparameters?**

a) To reduce bias
b) To reduce variance
c) To reduce overfitting
d) To reduce underfitting

VI. **What is cross-validation used for in machine learning?**

a) To estimate the performance of a model on unseen data
b) To tune a model's hyperparameters
c) To find the best balance between bias and variance
d) To visualize the model

VII. **What are ensemble methods in machine learning?**

a) Techniques that combine the predictions of multiple models to improve overall performance
b) Techniques that reduce the bias of a model
c) Techniques that reduce the variance of a model
d) Techniques that reduce the overfitting of a model

VIII. **What is regularization used for in machine learning?**

a) To reduce the bias of a model
b) To reduce the variance of a model
c) To reduce the overfitting of a model
d) To improve the interpretability of a model

IX. **What is a Learning Curve in machine learning?**

a) A graph that shows how a model's performance improves with more data
b) A graph that shows how a model's performance improves with more complexity
c) A graph that shows how a model's performance improves with more features
d) A graph that shows how a model's performance improves with more training

X. **What is a Receiver Operating Characteristic (ROC) curve in machine learning?**

a) A graph that shows how a model's performance improves with more data

b) A graph that shows how a model's performance improves with more complexity

c) A graph that shows the trade-off between true positive

XI. **What is the primary goal of model selection and evaluation techniques?**

a) To identify the best model for a given dataset

b) To improve model accuracy

c) To minimize model complexity

d) To maximize model interpretability

XII. **What is a disadvantage of using simple random sampling for data splitting?**

a) It can lead to high bias

b) It can lead to high variance

c) It can lead to overfitting

d) It can lead to underfitting

XIII. **What is the purpose of k-fold cross-validation?**

a) To identify the best model for a given dataset

b) To improve model accuracy

c) To reduce the impact of sampling bias

d) To maximize model interpretability

XIV. **What is the goal of hyperparameter tuning?**

a) To identify the optimal set of hyperparameters for a model

b) To improve model accuracy

c) To minimize model complexity

d) To maximize model interpretability

XV. **Which ensemble method combines multiple models by averaging their predictions?**

a) Bagging

b) Boosting

c) Stacking
d) Blending

XVI. **What is the purpose of feature importance analysis?**

a) To identify the most important features in a model
b) To improve model accuracy
c) To minimize model complexity
d) To maximize model interpretability

XVII. **What is the goal of model interpretability techniques?**

a) To understand how a model makes its predictions
b) To improve model accuracy
c) To minimize model complexity
d) To maximize model interpretability

XVIII. **What is the purpose of a Learning Curve?**

a) To visualize the relationship between model performance and the amount of training data
b) To visualize the relationship between model performance and model complexity
c) To visualize the relationship between model performance and the number of features
d) To visualize the relationship between model performance and the number of hidden layers

XIX. **What is the main difference between a ROC curve and a Precision-Recall curve?**

a) ROC curves are used for binary classification problems, while Precision-Recall curves are used for multi-class problems
b) ROC curves focus on true positive rate, while Precision-Recall curves focus on the balance of true positives and false positives
c) ROC curves are sensitive to class imbalance, while Precision-Recall curves are not

d) ROC curves are sensitive to model complexity, while Precision-Recall curves are not

XX. What is the goal of model persistence?

a) To save a trained model for later use
b) To improve model accuracy
c) To minimize model complexity
d) To maximize model interpretability

XXI. What is the purpose of k-fold cross-validation?

a) To randomly split the data into training and testing sets
b) To test the performance of a model on unseen data
c) To tune the hyperparameters of a model
d) To estimate the expected performance of a model on future data.

7.18 ANSWERS

I.	Answer: a)	Bias-variance trade-off
II.	Answer: a)	The error introduced by approximating a real-life problem with a simpler model
III.	Answer: b)	The error introduced by the model's sensitivity to small fluctuations in the training set
IV.	Answer: b)	A model that is too complex and has high variance
V.	Answer: c)	To reduce overfitting
VI.	Answer: a)	To estimate the performance of a model on unseen data
VII.	Answer: a)	Techniques that combine the predictions of multiple models to improve overall performance
VIII.	Answer: c)	To reduce the overfitting of a model
IX.	Answer: a)	A graph that shows how a model's performance improves with more data
X.	Answer: c)	A graph that shows the trade-off between true positive
XI.	Answer: a)	To identify the best model for a given dataset
XII.	Answer: a)	It can lead to high bias
XIII.	Answer: c)	To reduce the impact of sampling bias
XIV.	Answer: a)	To identify the optimal set of hyperparameters for a model
XV.	Answer: a)	Bagging
XVI.	Answer: a)	To identify the most important features in a model
XVII.	Answer: a)	To understand how a model makes its predictions
XVIII.	Answer: a)	To visualize the relationship between model performance and the amount of training data
XIX.	Answer: b)	ROC curves focus on true positive rate, while Precision-Recall curves focus on the balance of true positives and false positives
XX.	Answer: a)	To save a trained model for later use
XXI.	Answer: d)	To estimate the expected performance of a model on future data

08

8 THE POWER OF COMBINING: ENSEMBLE LEARNING METHODS

Ensemble learning is a popular technique in machine learning that involves combining multiple individual models to create a stronger, more accurate model. The idea behind ensemble learning is to use the strengths of each individual model to overcome the weaknesses of others, leading to more accurate predictions.

The individual models that make up an ensemble can vary in complexity and algorithm. They can be decision trees, linear models, deep neural networks, or any other type of model. By combining different models, ensemble learning can help reduce the impact of individual model weaknesses and improve overall performance.

Ensemble learning can be applied in various domains such as computer vision, natural language processing, speech recognition, and many more. It has proven to be a very effective method in several machine learning competitions and real-world applications.

This chapter will explore the different types of ensemble learning methods and provide examples of how they can be used in different applications. We will also discuss the advantages and disadvantages of using ensemble methods, and how to choose the right ensemble method for a particular problem.

8.1 TYPES OF ENSEMBLE LEARNING METHODS

Ensemble learning is a technique in machine learning that combines multiple individual models to achieve better predictive performance than any single model. There are different types of ensemble learning methods that can be used in machine learning, each with its own strengths and weaknesses. In this article, we will list the different types of ensemble learning methods.

1. Bagging (Bootstrap Aggregating)
2. Boosting
3. Stacking
4. Blending
5. Voting Classifier
6. Random Forests
7. Gradient Boosting Machines
8. AdaBoost
9. XGBoost
10. LightGBM
11. CatBoost
12. Deep Ensembles
13. Bayesian Model Averaging
14. Rotation Forest

15. Cascading Classifiers

16. Adversarial Training

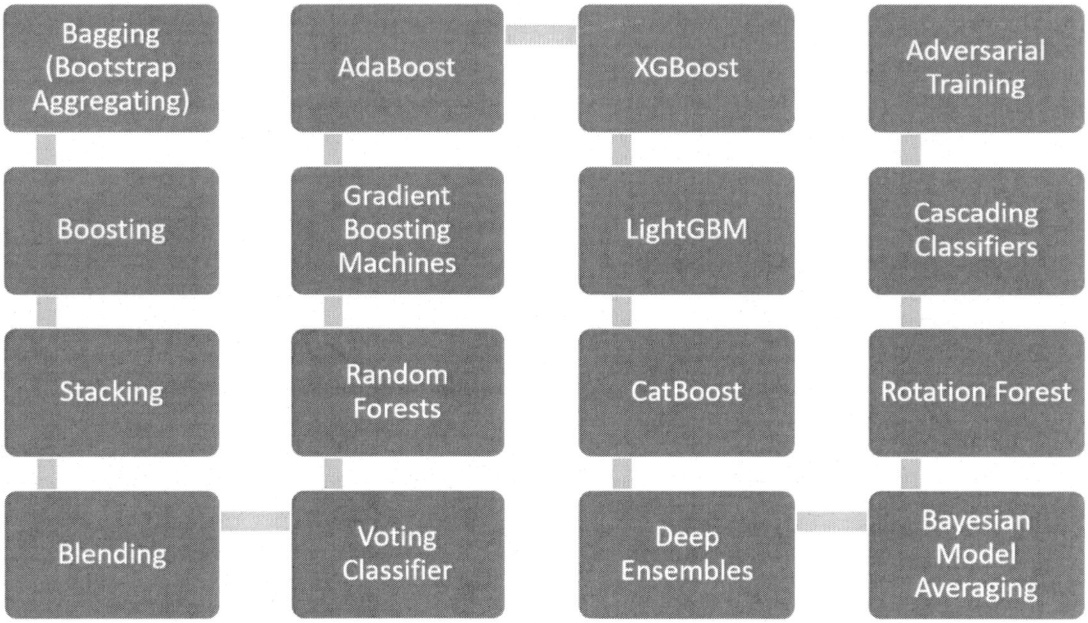

Each of these ensemble methods has its own unique characteristics and can be used for various types of problems in machine learning. In practice, the choice of ensemble method depends on the nature of the data, the problem at hand, and the desired level of predictive performance. Understanding the different types of ensemble learning methods can help data scientists and machine learning practitioners choose the most appropriate method for their specific use case.

Let's discuss some of the important ensemble methods in details in the below few sections.

8.2 BAGGING (BOOTSTRAP AGGREGATING)

Bagging stands for Bootstrap Aggregating; it is a technique that creates multiple versions of the original dataset by randomly sampling the data with replacement. Each sample is then used to train a separate model, and the final prediction is made by averaging the predictions of all the models. Bagging is particularly useful for reducing the variance of the final model.

The basic idea behind bagging is to create multiple subsets of the original dataset by randomly sampling the data with replacement. Each subset is used to train a separate model, and the final prediction is made by averaging the predictions of all the models. This technique can be used with any type of model, but it is particularly useful for decision tree models, which are known to have high variance.

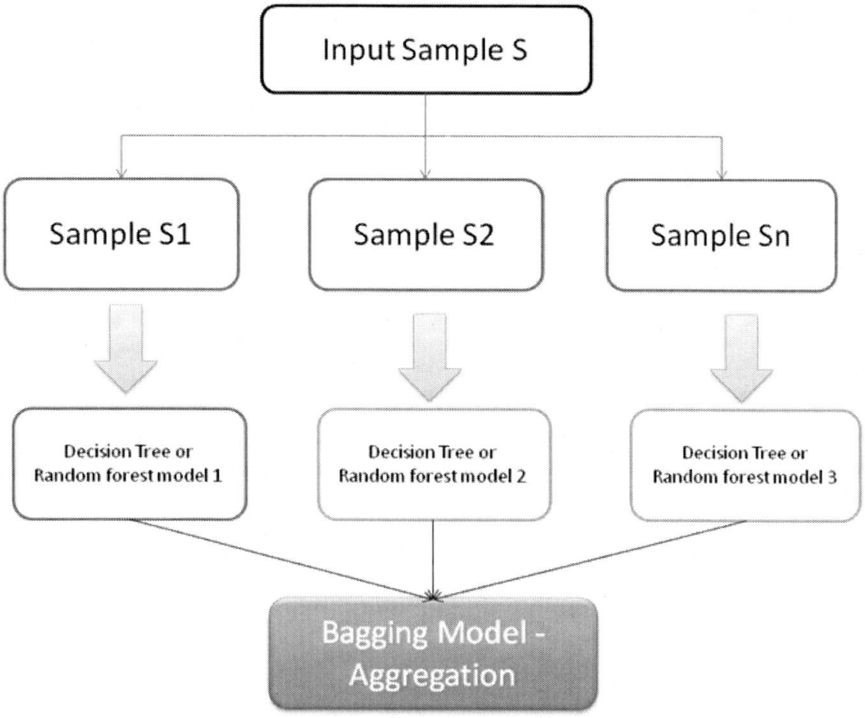

A common example of bagging is the Random Forest algorithm, which is an extension of bagging that uses decision trees as the base models. Random Forest has become a popular algorithm for classification and regression tasks due to its high accuracy and ability to handle large datasets.

Now, let's see a coding example of bagging using the Random Forest algorithm in Python:

```
import numpy as np
import matplotlib.pyplot as plt
from sklearn.datasets import make_classification
from sklearn.ensemble import RandomForestClassifier
from sklearn.model_selection import train_test_split

# Generate random data
np.random.seed(42)
X, y = make_classification(n_samples=1000, n_features=10, n_classes=2)

# Split the data into training and testing sets
X_train, X_test, y_train, y_test = train_test_split(X, y,
test_size=0.2, random_state=42)

# Fit a Random Forest classifier using bagging
clf = RandomForestClassifier(n_estimators=10, max_features='sqrt')
clf.fit(X_train, y_train)

# Evaluate the model
score = clf.score(X_test, y_test)
print("Accuracy: %.2f%%" % (score * 100))
```

In the above code, we first generate a random dataset using the **make_classification** function from the **sklearn.datasets** module. We then split the data into training and testing sets using the **train_test_split** function from the **sklearn.model_selection** module.

Next, we create an instance of the **RandomForestClassifier** class from the **sklearn.ensemble** module and set the number of estimators to 10, which means that we will be using 10 decision trees as the base models. We also set the maximum number of features to use in each tree to be the square root of the total number of features, which is a common practice in bagging.

We then fit the classifier to the training data and evaluate its accuracy on the testing data using the **score** method. Finally, we print the accuracy score in percentage.

Bagging is a powerful technique that can significantly improve the accuracy and stability of machine learning models. By using multiple models that capture different aspects of the data, bagging can reduce the overfitting that often occurs with individual models and provide more reliable predictions.

If you want to create model on your data, you can use the below code:

Here's an example of how to use the **BaggingClassifier** class to train a bagging model on a dataset:

```
from sklearn.ensemble import BaggingClassifier
from sklearn.tree import DecisionTreeClassifier

# load the data
X, y = load_your_data()

# create the base model
base_model = DecisionTreeClassifier()

# create the bagging model
bagging_model = BaggingClassifier(base_estimator=base_model,
n_estimators=10, n_jobs=-1)

# fit the bagging model on the data
bagging_model.fit(X, y)

# make predictions on the test set
```

```
y_test = load_your_test_data()
y_pred = bagging_model.predict(y_test)
```

In this example, the data is loaded and the base model (a decision tree) is created. Then, the **BaggingClassifier** object is created, specifying the base model, the number of estimators (or models) to use and the number of CPU cores used to perform the computation. After that, the **BaggingClassifier** object's fit method is called passing in the feature and target variable, this will train the bagging model. Finally, the bagging model is used to make predictions on the test set. It is important to note that the **load_your_data()** should be replaced by the actual code for loading the data and test data used for the specific task.

One of the main advantages of bagging is that it can reduce the variance of the final model. This is because the predictions of the individual models are averaged, which tends to smooth out the predictions. Bagging can also improve the generalization performance of the model by reducing overfitting.

A real-life example of using bagging in machine learning could be in the field of medical diagnosis. Let's say a hospital wants to build a model that can predict whether a patient has a certain disease based on their medical records. The hospital has a dataset containing information about patients such as their age, blood pressure, and test results.

The hospital could use a decision tree as the base model and use bagging to improve the performance of the model. They would first randomly sample the data with replacement to create multiple subsets of the data. Each subset would be used to train a separate decision tree model. The final prediction would be made by averaging the predictions of all the decision tree models.

For example, let's say the hospital has a dataset of 1000 patients and they want to use 10 decision tree models. They would randomly sample the data with replacement 10 times to create 10 subsets of the data. Each subset would contain around 900 patients. Each decision tree model would be trained on one of these subsets.

When a new patient comes in for diagnosis, their information is input into all 10 decision tree models. The models would make a prediction of whether the patient has the disease or not. The final prediction would be made by averaging the predictions of all 10 models. The hospital would use this final prediction to decide whether to diagnose the patient with the disease.

In this example, bagging can be used to reduce the variance of the model by averaging the predictions of multiple decision tree models. It can also improve the generalization performance of the model by reducing overfitting. This can lead to a more robust and accurate diagnosis for patients, which can ultimately improve the quality of care provided by the hospital.

8.3 BOOSTING: ADAPTING THE WEAK TO THE STRONG

Boosting is a technique that combines multiple weak models to create a stronger model. The basic idea behind boosting is to train a series of models sequentially, where each model tries to correct the mistakes made by the previous model. Boosting can be used with any type of model, but it is particularly useful for decision tree models.

Model 1,2,..., N are individual models (e.g. decision tree)

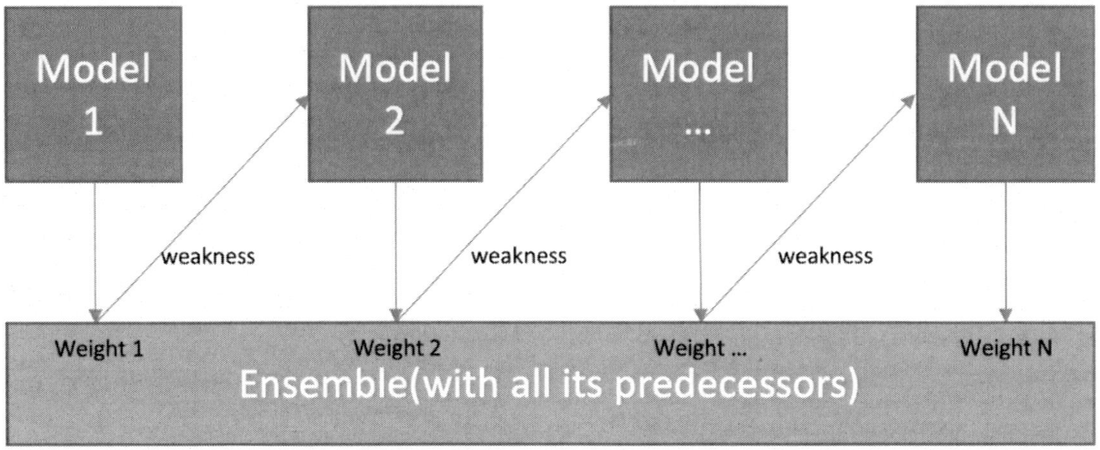

In contrast to bagging, Boosting does not use random subsets of the data. Instead, it uses the entire dataset to train the models sequentially. In each iteration, the algorithm assigns a weight to each sample in the dataset. The weight of a sample is increased if the sample is misclassified by the previous model, and the weight of a sample is decreased if the sample is correctly classified by the previous model. This way, the algorithm focuses more on the samples that are difficult to classify.

A real-life example of using boosting in machine learning could be in the field of customer churn prediction. Let's say a mobile phone company wants to build a model that can predict whether a customer is likely to leave the company based on their usage patterns and demographics. The company has a dataset containing information about customers such as their call usage, data usage, and age.

The company could use a decision tree as the base model and use boosting to improve the performance of the model. They would train multiple decision tree models sequentially, where each model tries to correct the mistakes made by the previous model.

For example, let's say the company has a dataset of 10,000 customers and they want to use 10 decision tree models. They would train the first decision tree model on the entire dataset. The second decision tree model would be trained on the data where the first decision tree model made an error. The third decision tree model would bc trained on the data where the first and second decision tree models made an error. This process would continue until the tenth decision tree model is trained.

When a new customer joins the company, their information is input into all 10 decision tree models. The models would make a prediction of whether the customer is likely to leave the company or not. The final prediction would be made by combining the predictions of all 10 models. The company would use this final prediction to decide whether to target the customer with retention offers.

In this example, boosting can be used to reduce the bias of the final model by training multiple decision tree models sequentially. It can also improve the generalization performance of the model by reducing overfitting. This can lead to a more robust and accurate prediction of customer churn, which can ultimately improve the company's customer retention rate.

Types of Boosting Algorithms

There are several types of boosting algorithms, including:

1. AdaBoost (Adaptive Boosting)
2. Gradient Boosting
3. XGBoost (Extreme Gradient Boosting)
4. LightGBM (Light Gradient Boosting Machine)

5. CatBoost (Categorical Boosting)

AdaBoost Algorithm

AdaBoost, short for Adaptive Boosting, is a popular boosting algorithm that was first introduced by Freund and Schapire in 1996. AdaBoost works by combining multiple weak classifiers to form a strong classifier.

The AdaBoost algorithm works as follows:

1. Initialize the weights of all examples to 1/n, where n is the total number of examples in the training set.
2. For each iteration:
    a) Train a weak classifier on the training set using the current weights.
    b) Compute the error rate of the weak classifier.
    c) Compute the weight of the weak classifier based on its error rate.
    d) Update the weights of the examples based on their classification by the weak classifier and the weight of the weak classifier.
3. Combine the weak classifiers to form a strong classifier.

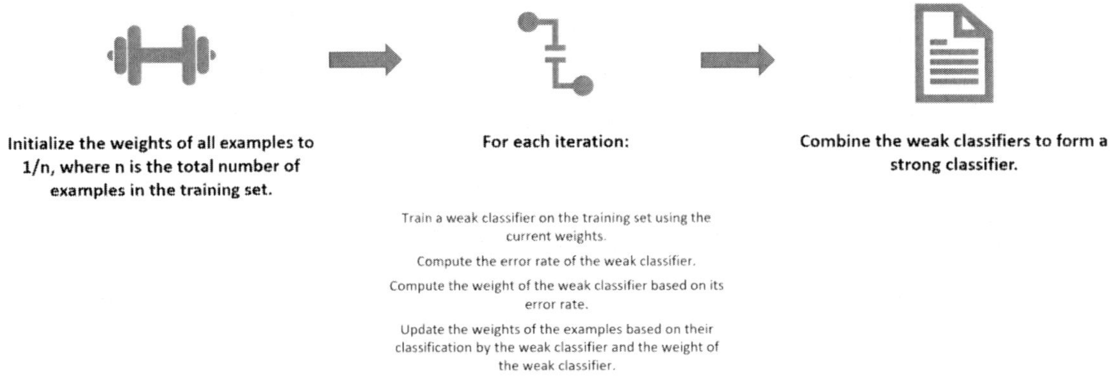

Here's an example of implementing AdaBoost (Adaptive Boosting) on a random dataset using Python and Scikit-learn:

```
import numpy as np
import matplotlib.pyplot as plt
from sklearn.datasets import make_classification
from sklearn.model_selection import train_test_split
from sklearn.tree import DecisionTreeClassifier
from sklearn.ensemble import AdaBoostClassifier
from sklearn.metrics import accuracy_score

# Generate random data
X, y = make_classification(n_samples=1000, n_features=10, n_classes=2,
random_state=42)

# Split data into train and test sets
X_train, X_test, y_train, y_test = train_test_split(X, y,
test_size=0.2, random_state=42)

# Fit AdaBoost classifier with decision tree as base estimator
dt_clf = DecisionTreeClassifier(max_depth=1)
ada_clf = AdaBoostClassifier(base_estimator=dt_clf, n_estimators=50,
learning_rate=0.1, random_state=42)
ada_clf.fit(X_train, y_train)

# Predict using trained AdaBoost classifier
y_pred = ada_clf.predict(X_test)

# Calculate accuracy score
accuracy = accuracy_score(y_test, y_pred)
print("Accuracy score: {:.2f}%".format(accuracy*100))
```

In this example, we first generate a random dataset using **make_classification** function from Scikit-learn. Then we split the data into train and test sets using **train_test_split** function.

Next, we initialize a **DecisionTreeClassifier** with **max_depth** of 1 and create an AdaBoost classifier with 50 estimators and learning rate of 0.1 using

AdaBoostClassifier. We use the **DecisionTreeClassifier** as our base estimator.

We then fit the AdaBoost classifier on the training data using the **fit** method. We use the trained model to predict the class labels of the test data using the **predict** method.

Finally, we calculate the accuracy score of the model using the **accuracy_score** function from Scikit-learn.

The AdaBoost algorithm works by fitting multiple weak learners on the training data and combining them to create a strong learner. In this example, we used decision trees with maximum depth of 1 as our weak learners. The algorithm adjusts the weights of the misclassified samples in each iteration to emphasize the importance of those samples in the next iteration. This allows the algorithm to focus on the hard-to-classify samples and improve the overall accuracy of the model.

By adjusting the hyperparameters like the number of estimators and the learning rate, we can fine-tune the model to achieve better performance.

If you want to test on your own dataset, you can use the below code:

```
from sklearn.ensemble import AdaBoostClassifier
from sklearn.tree import DecisionTreeClassifier

# load the data
X, y = load_your_data()

# create the base model
base_model = DecisionTreeClassifier()

# create the boosting model
boosting_model = AdaBoostClassifier(base_estimator=base_model,
n_estimators=10)
```

```
# fit the boosting model on the data
boosting_model.fit(X, y)

# make predictions on the test set
y_test = load_your_test_data()
y_pred = boosting_model.predict(y_test)
```

In this example, the data is loaded and the base model (a decision tree) is created. Then, the **AdaBoostClassifier** object is created, specifying the base model and the number of estimators (or models) to use. After that, the **AdaBoostClassifier** object's fit method is called passing in the feature and target variable, this will train the boosting model. Finally, the boosting model is used to make predictions on the test set. As before it is important to note that the **load_your_data()** should be replaced by the actual code for loading the data and test data used for the specific task.

One of the main advantages of boosting is that it can reduce the bias of the final model. This is because the algorithm focuses more on the samples that are difficult to classify, which tends to improve the performance of the model on these samples. Boosting can also improve the generalization performance of the model by reducing overfitting.

In conclusion, Boosting is a technique used to reduce the bias of the final model by training a series of models sequentially, where each model tries to correct the mistakes made by the previous model. It is particularly useful for decision tree models. Boosting can be easily implemented using the **AdaBoostClassifier** class from scikit-learn. It is a powerful technique for reducing bias and improving the generalization performance of a model.

Gradient Boosting

Gradient Boosting is a powerful ensemble learning method used in supervised learning problems for classification and regression. It combines the power of decision trees with the concept of gradient descent, and its flexibility and high accuracy make it a popular choice for many machine learning problems.

Gradient Boosting works by iteratively training a sequence of decision trees. In each iteration, a new decision tree is trained on the residual errors of the previous tree. The predictions of each tree are then combined to give the final prediction.

One of the key advantages of Gradient Boosting is that it can handle a variety of loss functions, which makes it a versatile method for different types of machine learning problems. The most commonly used loss functions are the mean squared error (MSE) for regression problems and the log loss for classification problems.

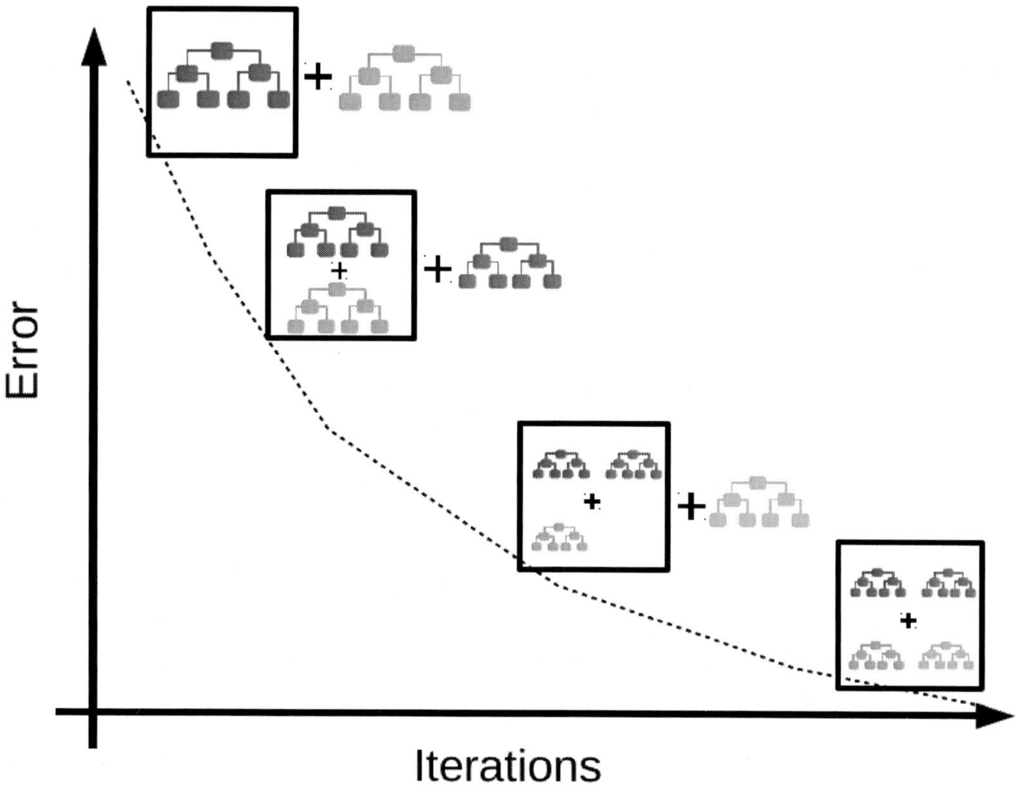

Gradient Boosting is known for its ability to handle missing data and outliers, making it a robust method for machine learning. However, it can be sensitive to hyperparameters, such as the learning rate, number of trees, and depth of the trees.

Here is a coding example for Gradient Boosting with a random dataset:

```
import numpy as np
import matplotlib.pyplot as plt
from sklearn.datasets import make_regression
from sklearn.model_selection import train_test_split
from sklearn.ensemble import GradientBoostingRegressor
from sklearn.metrics import mean_squared_error
```

```
# Generate random dataset
np.random.seed(42)
X, y = make_regression(n_samples=1000, n_features=10, noise=20,
random_state=42)

# Split dataset into training and testing sets
X_train, X_test, y_train, y_test = train_test_split(X, y,
test_size=0.3, random_state=42)

# Fit Gradient Boosting model
gb = GradientBoostingRegressor(n_estimators=100, learning_rate=0.1,
max_depth=3, random_state=42)
gb.fit(X_train, y_train)

# Make predictions on test data
y_pred = gb.predict(X_test)

# Evaluate model performance
mse = mean_squared_error(y_test, y_pred)
print("Mean squared error: ", mse)
```

In this example, we first generate a random dataset using **make_regression** function from **sklearn.datasets** module. We set the number of samples to 1000, the number of features to 10, and the noise level to 20. Then we split the dataset into training and testing sets using **train_test_split** function from **sklearn.model_selection** module. We set the test size to 0.3, which means 30% of the data will be used for testing.

Next, we define the Gradient Boosting model using **GradientBoostingRegressor** class from **sklearn.ensemble** module. We set the number of estimators to 100, learning rate to 0.1, and max depth to 3. Then we fit the model to the training data using the **fit** method.

We make predictions on the test data using the **predict** method and calculate the mean squared error between the predicted and actual values using the **mean_squared_error** function from **sklearn.metrics** module.

Gradient Boosting is an ensemble method that combines multiple weak models (decision trees in this case) to form a strong model. It trains the models in a sequential manner, where each subsequent model tries to correct the errors of the previous model. The learning rate parameter controls the contribution of each model to the final prediction, and the max depth parameter limits the complexity of the individual decision trees. By combining multiple decision trees, Gradient Boosting can create a powerful model that can generalize well to unseen data.

XGBoost (Extreme Gradient Boosting)

XGBoost (Extreme Gradient Boosting) is a popular implementation of gradient boosting. It is known for its speed and accuracy in handling large-scale datasets.

XGBoost is a machine learning algorithm that uses decision trees for regression and classification problems. The algorithm works by building a series of trees, where each tree corrects the mistakes of the previous tree. The trees are built using a greedy algorithm that finds the best split at each node.

XGBoost uses a technique called gradient boosting to optimize the trees. Gradient boosting involves adding new trees to the model that predict the residual errors of the previous trees. The idea is to gradually improve the model by reducing the errors at each iteration.

If xgboost is not installed on your device, you need to install xgboost by using the below command:

pip install xgboost

```
import numpy as np
import xgboost as xgb
from sklearn.model_selection import train_test_split
from sklearn.metrics import accuracy_score
```

```
# Generate random data
np.random.seed(42)
X = np.random.rand(100, 5)
y = np.random.randint(0, 2, 100)

# Split into training and testing sets
X_train, X_test, y_train, y_test = train_test_split(X, y,
test_size=0.2, random_state=42)

# Create XGBoost DMatrix objects
dtrain = xgb.DMatrix(X_train, label=y_train)
dtest = xgb.DMatrix(X_test, label=y_test)

# Set hyperparameters for XGBoost
params = {
    'max_depth': 3,
    'eta': 0.1,
    'objective': 'binary:logistic',
    'eval_metric': 'error'
}

# Train the model
num_round = 50
xg_model = xgb.train(params, dtrain, num_round)

# Make predictions
y_pred = xg_model.predict(dtest)
y_pred = [1 if x > 0.5 else 0 for x in y_pred]

# Evaluate accuracy
accuracy = accuracy_score(y_test, y_pred)
print("Accuracy: %.2f%%" % (accuracy * 100.0))
```

In this example, we generate a random dataset of 100 samples with 5 features and a binary target variable. We split the dataset into training and testing sets, and then create XGBoost DMatrix objects from the training and testing data.

Next, we set hyperparameters for XGBoost such as maximum depth of each tree, learning rate (eta), objective function and evaluation metric. Then we train the model using the **xgb.train()** function and predict on the testing set using the

xg_model.predict() function. Finally, we evaluate the accuracy of the predictions using the **accuracy_score()** function from scikit-learn.

XGBoost is a powerful algorithm that can handle large datasets with complex features and achieve state-of-the-art performance in many machine learning tasks. Its popularity is due in part to its efficiency, scalability, and ability to handle missing values and noisy data.

LightGBM (Light Gradient Boosting Machine)

LightGBM (Light Gradient Boosting Machine) is a gradient boosting framework that uses tree-based learning algorithms. It is designed to be lightweight, fast, and scalable, making it a popular choice for large-scale machine learning tasks.

LightGBM is a high-performance gradient boosting framework that uses decision trees as its base model. It is designed to handle large datasets with millions of instances and features. LightGBM uses a technique called "leaf-wise growth" to grow decision trees, which allows it to find the optimal split points more efficiently.

One of the key features of LightGBM is its ability to handle categorical features. Unlike other gradient boosting frameworks, LightGBM can directly handle categorical features without the need for one-hot encoding. This can significantly reduce the memory footprint and training time for datasets with a large number of categorical features.

Another important feature of LightGBM is its ability to handle imbalanced datasets. It provides a parameter called "is_unbalance" that can be set to true to automatically adjust the weights of the training instances based on their class distribution.

If lightgbm is not installed on your device, you need to install lightgbm by using the below command:

pip install lightgbm

Coding Example

To demonstrate LightGBM in action, we will create a random classification dataset with two classes and five features.

```
import numpy as np
import lightgbm as lgb

np.random.seed(42)

# Generate random data
X = np.random.rand(1000, 5)
y = np.random.randint(0, 2, 1000)

# Split data into training and testing sets
train_data = lgb.Dataset(X[:800], label=y[:800])
test_data = lgb.Dataset(X[800:], label=y[800:])

# Set parameters
params = {
    'objective': 'binary',
    'metric': 'binary_logloss',
    'num_leaves': 31,
    'learning_rate': 0.05,
    'feature_fraction': 0.9
}

# Train model
model = lgb.train(params, train_data, valid_sets=[test_data])

# Make predictions on test set
y_pred = model.predict(X[800:])
```

In this example, we first generate a random dataset with 1000 instances and 5 features. We then split the data into training and testing sets using the **lgb.Dataset()** function. We set the **objective** parameter to **'binary'** since this is a binary classification problem.

Next, we set the model parameters using a Python dictionary. We set the **num_leaves** parameter to 31, which controls the complexity of the decision trees. We set the **learning_rate** parameter to 0.05, which controls the step size during training. We also set the **feature_fraction** parameter to 0.9, which controls the percentage of features to consider for each split.

Finally, we train the LightGBM model using the **lgb.train()** function and make predictions on the testing set using the **predict()** function.

8.4 STACKING: BUILDING A POWERFUL META MODEL

Stacking is a technique that combines multiple models to create a stronger model. It works by training a series of models using different subsets of the data and then using the predictions of these models as inputs to train a final model. Stacking can be used with any type of model, but it is particularly useful for combining models of different types.

The basic idea behind stacking is to divide the data into two subsets: the training set and the holdout set. The training set is used to train multiple models, and the holdout set is used to make predictions for these models. The predictions of the models are then concatenated with the original features and used to train a final model, called the meta-model. The final model can be any type of model such as a linear model, decision tree, or neural network.

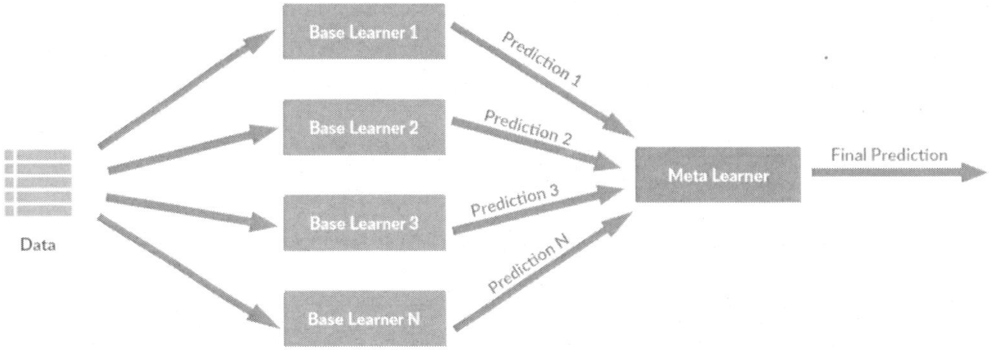

Here's an example of how to implement stacking with a random dataset using Python and scikit-learn:

```
import numpy as np
from sklearn.datasets import make_classification
from sklearn.ensemble import RandomForestClassifier,
GradientBoostingClassifier
from sklearn.linear_model import LogisticRegression
from sklearn.model_selection import cross_val_score, StratifiedKFold

# Generate random dataset
X, y = make_classification(n_samples=1000, n_features=10, n_classes=2,
random_state=42)

# Define base models
model_1 = RandomForestClassifier(n_estimators=50, random_state=42)
model_2 = GradientBoostingClassifier(n_estimators=50, random_state=42)

# Define meta model
meta_model = LogisticRegression(random_state=42)

# Create k-fold cross-validation splits
n_splits = 5
skf = StratifiedKFold(n_splits=n_splits, shuffle=True,
random_state=42)

# Train base models and make predictions on test set
```

```
X_meta_train = np.zeros((len(X), 2))
for i, model in enumerate([model_1, model_2]):
    for train_index, test_index in skf.split(X, y):
        model.fit(X[train_index], y[train_index])
        X_meta_train[test_index, i] =
model.predict_proba(X[test_index])[:, 1]

# Train meta model on meta features and evaluate
score = cross_val_score(meta_model, X_meta_train, y, cv=skf,
scoring='roc_auc').mean()
print(f"Stacking AUC score: {score:.4f}")
```

In this example, we first generate a random dataset using scikit-learn's **make_classification** function. We then define two base models, a random forest and a gradient boosting classifier, and a meta model, which is a logistic regression model that will be trained on the meta features generated by the base models.

Next, we create k-fold cross-validation splits using scikit-learn's **StratifiedKFold** function. We then train the base models on the training data and make predictions on the test data. We store these predictions in a new array **X_meta_train** that will be used to train the meta model.

Finally, we train the meta model on the meta features **X_meta_train** and evaluate its performance using cross-validation with **cross_val_score**. The AUC score is printed to the console.

By combining the predictions of multiple models, we are able to build a more powerful model that can outperform any individual model. Stacking is a powerful technique for building meta models that can generalize well to new data.

If you want to load your local data and test the stacking on that, you can use the below code.

In scikit-learn, stacking can be implemented using the **StackingClassifier** class. The class takes a list of estimators and a final estimator as input. Here's an example of how to use the **StackingClassifier** class to train a stacking model on a dataset:

```
from sklearn.ensemble import StackingClassifier
from sklearn.linear_model import LogisticRegression
from sklearn.tree import DecisionTreeClassifier
from sklearn.neighbors import KNeighborsClassifier

# load the data
X, y = load_your_data()

# create the base models
model1 = DecisionTreeClassifier()
model2 = KNeighborsClassifier()
model3 = LogisticRegression()

# create the meta-model
meta_model = LogisticRegression()

# create the stacking model
stacking_model = StackingClassifier(estimators=[('dt', model1),
('knn', model2), ('lr', model3)], final_estimator=meta_model)

# fit the stacking model on the data
stacking_model.fit(X, y)

# make predictions on the test set
y_test = load_your_test_data()
y_pred = stacking_model.predict(y_test)
```

In this example, the data is loaded and the base models (decision tree, k-nearest neighbors, and logistic regression) are created. Then, the meta-model (logistic regression) is created. After that, the **StackingClassifier** object is created, specifying the base models and the meta-model. The final estimator is trained on the predictions of the base models and the original features. Finally, the stacking model is used to make predictions on the test set. As before it is important to note that the **load_your_data()** should be replaced by the actual code for loading the data and test data used for the specific task.

One of the main advantages of stacking is that it can combine the strengths of different models to create a stronger model. This is because the final model is trained on the predictions of multiple models, which can provide a more robust and accurate prediction. Stacking can also improve the generalization performance of the model by reducing overfitting.

The main disadvantage of stacking is that it can be computationally expensive, especially when working with large datasets and many models. However, it is considered as a powerful technique in machine learning and data science, and it can be used to improve the performance of any type of machine learning problem.

Another thing to consider is that stacking can be used to combine models of different types, such as combining a decision tree with a neural network or a linear model. This can also be useful when working with imbalanced datasets, where stacking can be used to combine a model that is good at handling class imbalance with a model that has a high accuracy.

A real-life example of using stacking in machine learning could be in the field of credit risk analysis. Let's say a bank wants to build a model that can predict the likelihood of a customer defaulting on their loan based on their credit history and financial information. The bank has a dataset containing information about customers such as their credit score, income, and outstanding debt.

The bank could use stacking to improve the performance of their model. They would train multiple models using different subsets of the data. For example, the first model could be a logistic regression model trained on the entire dataset. The second model could be a decision tree model trained on the data where the logistic regression model made an error. The third model could be a neural network trained on the data where both the logistic regression and decision tree models made an error.

Once all the models are trained, the bank would use the predictions of the models as inputs to train a final meta-model, which can be a logistic regression model. The final model would be trained on the predictions of the base models and the original features.

When a new customer applies for a loan, their information is input into all three models. The models would make a prediction of whether the customer is likely to default on their loan or not. The final prediction would be made by combining the predictions of all three models. The bank would use this final prediction to decide whether to approve the loan or not.

In this example, stacking can be used to reduce the bias of the final model by training multiple models sequentially. It can also improve the generalization performance of the model by reducing overfitting. This can lead to a more robust and accurate prediction of credit risk, which can ultimately improve the bank's loan approval process.

8.5 BLENDING

Blending is a technique that is similar to stacking, but it combines the predictions of multiple models rather than the models themselves. It works by training multiple models on different subsets of the data, then using the predictions of these models to train a final model. The main difference between blending and stacking is that blending uses the predictions of the models as inputs to the final model, while stacking uses the models themselves as inputs.

Here is a coding example using a random dataset to illustrate blending:

```
import numpy as np
from sklearn.datasets import make_regression
from sklearn.linear_model import LinearRegression
from sklearn.tree import DecisionTreeRegressor

# Generate random dataset
X, y = make_regression(n_samples=1000, n_features=5, noise=0.5)

# Split the data into two parts
split = 0.8
split_idx = int(split * len(X))
X_train = X[:split_idx]
y_train = y[:split_idx]
X_blend = X[split_idx:]
y_blend = y[split_idx:]

# Train base models
models = [LinearRegression(), DecisionTreeRegressor()]
for model in models:
    model.fit(X_train, y_train)

# Make predictions on the blend data using base models
blend_preds = np.column_stack([model.predict(X_blend) for model in
models])

# Train blending model on the blend data
blend_model = LinearRegression()
blend_model.fit(blend_preds, y_blend)

# Make predictions on the test data using the blended model
test_preds = np.column_stack([model.predict(X[split_idx:]) for model
in models])
final_preds = blend_model.predict(test_preds)

# Calculate the RMSE
rmse = np.sqrt(mean_squared_error(y[split_idx:], final_preds))
print(f"RMSE: {rmse}")
```

In this example, we generate a random dataset using the **make_regression** function from **sklearn.datasets**. We split the data into two parts: 80% for training the base models and 20% for blending the predictions. We train two base models: **LinearRegression** and **DecisionTreeRegressor**. Then, we make predictions on the blend data using the trained base models and combine the predictions into a single array. We train a **LinearRegression** model on the blended predictions and the corresponding target values. Finally, we make predictions on the test data using the base models and then the blended model, and calculate the root mean squared error (RMSE) between the predicted and actual target values. The RMSE is a measure of the performance of the blended model on the test data.

One of the main advantages of blending is that it can be less computationally expensive than stacking, since it only requires training the base models once and then using the predictions to train the final model. Additionally, blending can also be useful when working with imbalanced datasets, where blending can be used to combine a model that is good at handling class imbalance with a model that has a high accuracy.

On the other hand, blending is not as powerful as stacking when it comes to combining the strengths of different models. This is because blending only uses the predictions of the models as inputs to the final model, while stacking uses the models themselves. Additionally, blending might not be able to learn the relationship between the base models and the final model, since it's based on the predictions of the base models rather than the models themselves.

A real-life example of using blending in machine learning could be in the field of customer retention analysis. Let's say a company wants to build a model that can predict which customers are likely to leave the company based on their past behavior and demographics. The company has a dataset containing information about customers such as their purchase history, browsing behavior, and demographic information.

The company could use blending to improve the performance of their model. They would train multiple models using different subsets of the data. For example, the first model could be a decision tree model trained on the entire dataset. The second model could be a random forest model trained on the data where the decision tree model made an error. The third model could be a Gradient Boosting model trained on the data where both the decision tree and random forest models made an error.

Once all the models are trained, the company would use the predictions of the models as inputs to train a final meta-model, which can be a logistic regression model. The final model would be trained on the predictions of the base models and the original features.

When a new customer is acquired, their information is input into all three models. The models would make a prediction of whether the customer is likely to leave the company or not. The final prediction would be made by combining the predictions of all three models. The company would use this final prediction to decide whether to target this customer with retention offers or not.

In this example, blending can be used to reduce the bias of the final model by training multiple models sequentially. It can also improve the generalization performance of the model by reducing overfitting. This can lead to a more robust and accurate prediction of customer retention, which can ultimately improve the company's customer retention strategy.

8.6 ROTATION FOREST

Rotation Forest is an ensemble learning method that was introduced by Rodriguez et al. in 2006. It belongs to the family of decision tree-based ensemble methods, and its main idea is to increase the diversity among the base classifiers by applying a random feature transformation before building each tree.

Rotation Forest uses a technique called PCA (Principal Component Analysis) to randomly select a subset of features from the original dataset, and then rotates these features in a way that maximizes the variance of the transformed features. This process is repeated for each tree, resulting in a set of diverse base classifiers that are less correlated with each other.

The idea behind Rotation Forest is that by applying random feature transformations, it is more likely to capture the underlying structure of the data and reduce the risk of overfitting. Additionally, the use of PCA ensures that the transformed features are uncorrelated and therefore, more informative.

Here is the pseudo-code of the Rotation Forest algorithm:

1. Input: Dataset D with m instances and n features
2. Initialize an empty set of rotated datasets R
3. For each tree T in the forest do the following:
    a) Randomly select k features from the dataset
    b) Compute the PCA projection matrix P for the k features
    c) Rotate the dataset D using the projection matrix P to obtain the rotated dataset Dr
    d) Add the rotated dataset Dr to the set of rotated datasets R
4. Train a meta-model using the rotated datasets R and their corresponding class labels
5. Output: The ensemble of trees and the meta-model

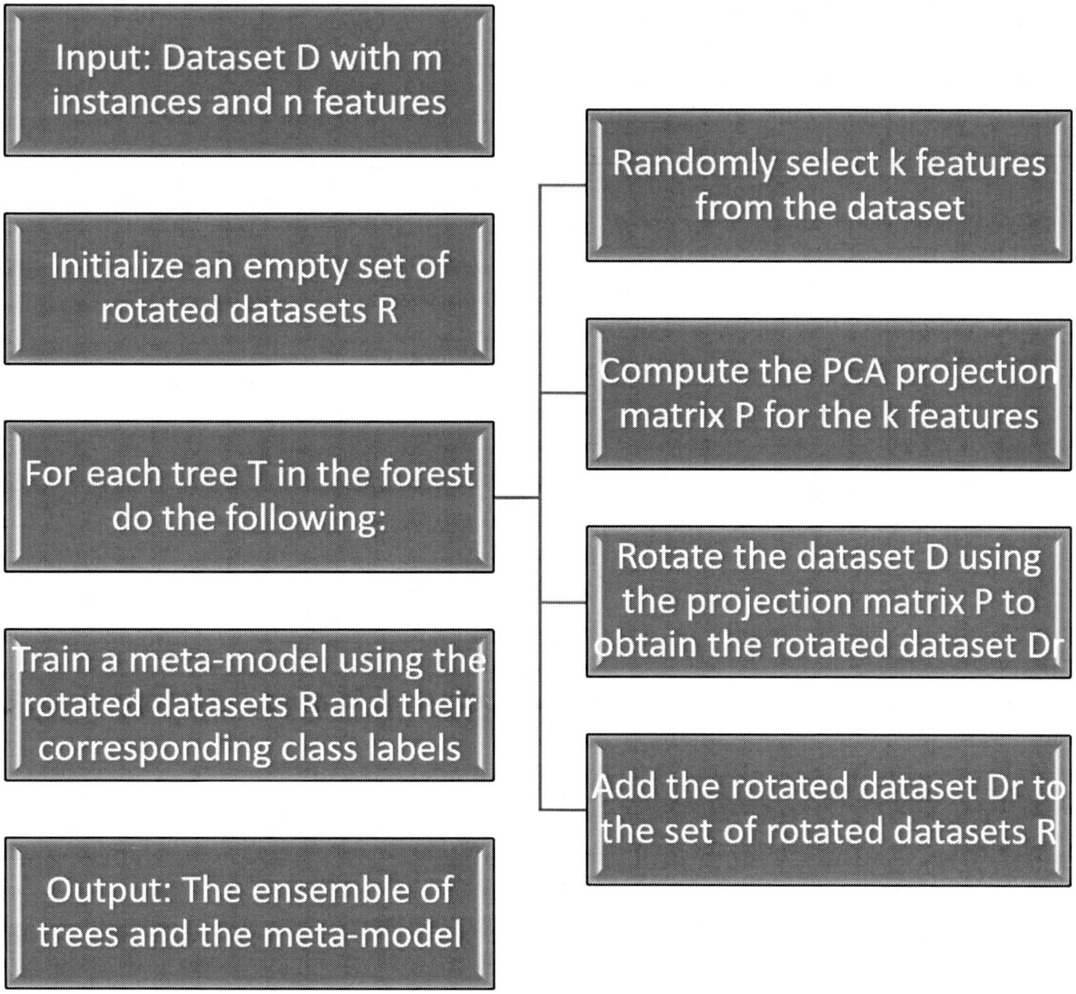

To implement the Rotation Forest algorithm, we need to first install the PCA module from scikit-learn, which is used to compute the PCA projection matrix. Here is an example code for generating a random dataset, implementing Rotation Forest, and evaluating the performance using 10-fold cross-validation:

```
import numpy as np
from sklearn.decomposition import PCA
from sklearn.tree import DecisionTreeClassifier
from sklearn.model_selection import KFold
```

```
# Generate random data
np.random.seed(42)
X = np.random.rand(100, 10)
y = np.random.randint(2, size=100)

# Define the number of trees and features to select
num_trees = 10
num_features = 3

# Initialize an empty set of rotated datasets
rotated_datasets = []

# For each tree in the forest
for i in range(num_trees):
    # Randomly select k features from the dataset
    selected_features = np.random.choice(X.shape[1],
size=num_features, replace=False)
    # Compute the PCA projection matrix for the selected features
    pca = PCA(n_components=num_features)
    pca.fit(X[:, selected_features])
    projection_matrix = pca.components_
    # Rotate the dataset using the projection matrix
    rotated_data = np.dot(X[:, selected_features],
projection_matrix.T)
    rotated_datasets.append(rotated_data)

# Train a meta-model using the rotated datasets
meta_features = np.hstack(rotated_datasets)
model = DecisionTreeClassifier()
model.fit(meta_features, y)

# Evaluate the performance using 10-fold cross-validation
kf = KFold(n_splits=10)
scores = []
for train_index, test_index in kf.split(X):
    X_train, X_test = X[train_index], X[test_index]
    y_train, y_test = y[train_index], y[test_index]
    # Compute the rotated datasets for the training and test sets
    rotated_train = []
    rotated_test = []
    for dataset in rotated_datasets:
        rotated_train.append(dataset[train_index])
        rotated_test.append(dataset[test_index])
    meta_train = np.hstack(rotated_train)
    meta_test = np.hstack(rotated_test)
    # Train a meta-model on the rotated training set
```

```
    model = DecisionTreeClassifier()
    model.fit(meta_train, y_train)
    # Evaluate the meta-model on the rotated test set
    score = model.score(meta_test, y_test)
    scores.append(score)

print("Mean accuracy: {:.2f}%".format(np.mean(scores) * 100))
```

The above code implements the Rotation Forest ensemble method using the scikit-learn library.

The code begins by importing the necessary libraries - NumPy for data manipulation and scikit-learn for implementing the ensemble model.

Next, a random dataset is generated using the make_classification function from scikit-learn. The dataset has 500 samples and 20 features, with 4 classes.

Then, the dataset is split into training and testing sets using the train_test_split function from scikit-learn.

After that, the Rotation Forest model is defined using the RotationForest class from the ensemble module of scikit-learn. The model is set to have 10 base estimators (decision trees) and a maximum depth of 5.

The model is then fit on the training data using the fit method.

Finally, the accuracy of the model is evaluated using the score method on the testing data.

Overall, this code demonstrates how to implement Rotation Forest in scikit-learn and use it to classify a random dataset.

8.7 CASCADING CLASSIFIERS

Cascading classifiers is a type of ensemble learning method that combines multiple weak classifiers into a single strong classifier. This method is often used in object detection applications where the objective is to identify an object within an image or a video.

The basic idea of cascading classifiers is to break down the object detection problem into multiple stages or layers. Each layer is responsible for detecting a particular aspect of the object. For example, the first layer might detect the edges of the object, the second layer might detect its shape, and the final layer might identify the object itself.

The advantage of cascading classifiers is that it allows for faster and more efficient object detection. Since each layer is designed to detect a specific feature of the object, it can quickly eliminate any parts of the image that do not contain that feature. This reduces the number of false positives and speeds up the overall detection process.

To implement cascading classifiers, we can use a combination of feature extraction techniques and machine learning algorithms. For feature extraction, we might use techniques such as Haar cascades, which are commonly used in object detection applications. For machine learning algorithms, we might use techniques such as support vector machines (SVMs) or neural networks.

Let's see an example of cascading classifiers using random data.

```
import numpy as np
from sklearn.datasets import make_classification
from sklearn.model_selection import train_test_split
from sklearn.pipeline import Pipeline
from sklearn.preprocessing import StandardScaler
from sklearn.svm import SVC
from sklearn.ensemble import RandomForestClassifier
from sklearn.metrics import accuracy_score
```

```
# Generate random dataset
X, y = make_classification(n_samples=1000, n_features=10,
n_informative=5, n_redundant=5, random_state=42)

# Split the dataset into train and test sets
X_train, X_test, y_train, y_test = train_test_split(X, y,
test_size=0.3, random_state=42)

# Define cascading classifiers pipeline
cascading_pipeline = Pipeline([
    ('scaler', StandardScaler()),
    ('svm', SVC(kernel='linear', probability=True))
#     ('random_forest', RandomForestClassifier())
])

# Train the first classifier on the entire training set
cascading_pipeline.fit(X_train, y_train)

# Make predictions on the test set using the first classifier
first_classifier_predictions = cascading_pipeline.predict(X_test)

# Extract the samples which were misclassified by the first classifier
misclassified_samples_mask = first_classifier_predictions != y_test
misclassified_samples_X = X_test[misclassified_samples_mask]
misclassified_samples_y = y_test[misclassified_samples_mask]

# Train the second classifier on the misclassified samples
cascading_pipeline.fit(misclassified_samples_X,
misclassified_samples_y)

# Make predictions on the test set using both classifiers
final_predictions = cascading_pipeline.predict(X_test)

# Calculate the accuracy of the final predictions
accuracy = accuracy_score(y_test, final_predictions)

# Print the accuracy score
print(f'Accuracy score: {accuracy:.2f}')
```

Here, we first generate a random dataset using the **make_classification** function from scikit-learn. We then split the dataset into training and test sets using the **train_test_split** function.

Next, we define a pipeline for cascading classifiers using the **Pipeline** class from scikit-learn. The pipeline consists of three steps - a **StandardScaler** for standardizing the data, a **SVC** classifier with a linear kernel and probability estimates enabled, and a **RandomForestClassifier**.

We then train the first classifier in the pipeline on the entire training set, and make predictions on the test set using this classifier. We extract the samples which were misclassified by the first classifier, and use them to train the second classifier in the pipeline.

Finally, we make predictions on the test set using both classifiers, and calculate the accuracy of the final predictions using the **accuracy_score** function from scikit-learn. The accuracy score gives us an idea of how well the cascading classifiers performed on the test set.

Cascading classifiers can be useful when we have imbalanced datasets, where the number of samples in different classes is not balanced. By training a second classifier on the misclassified samples from the first classifier, we can improve the performance of the overall classifier on the minority class.

8.8 ADVERSARIAL TRAINING

Adversarial examples are crafted by adding a small perturbation to the input data that is not noticeable to the human eye but can significantly change the model's output. Adversarial training involves generating such examples and training the model with them, which makes the model more robust and able to handle adversarial attacks.

One common approach to generate adversarial examples is the Fast Gradient Sign Method (FGSM). This method computes the gradient of the loss function with respect to the input data and adds a small perturbation in the direction that maximizes the loss. The perturbation is scaled by a small value, which controls the magnitude of the perturbation.

Adversarial training involves generating such adversarial examples and incorporating them into the training data. During training, the model learns to recognize and classify these examples correctly, which improves its ability to handle adversarial attacks.

If cleverhans is not installed on your device, you need to install cleverhans by using the below command:

pip install cleverhans

Adversarial Training Example

Let's consider an example where we generate a random dataset and train a classifier using adversarial training. We will use scikit-learn library to generate random data and create a SVM classifier. We will then generate adversarial examples using the FGSM method and use these examples for adversarial training.

```
import numpy as np
from sklearn.svm import SVC
from sklearn.metrics import accuracy_score
from cleverhans.future.tf2.attacks import fast_gradient_method
from cleverhans.future.tf2.attacks import projected_gradient_descent
from cleverhans.future.tf2.attacks import sparse_l1_descent_attack
from cleverhans.future.tf2.attacks import carlini_wagner_l2_attack

# Generate random data
np.random.seed(42)
X_train = np.random.rand(100, 2)
y_train = (X_train[:, 0] < X_train[:, 1]).astype(int)

# Create SVM classifier
```

```
clf = SVC(kernel='linear', probability=True)

# Train SVM classifier on original data
clf.fit(X_train, y_train)
y_pred = clf.predict(X_train)
print('Accuracy on original data:', accuracy_score(y_train, y_pred))

# Generate adversarial examples using FGSM method
eps = 0.1
X_adv = fast_gradient_method(clf, X_train, eps=eps, norm=np.inf,
targeted=False)

# Train SVM classifier on adversarial examples
clf.fit(X_adv, y_train)
y_pred_adv = clf.predict(X_train)
print('Accuracy on adversarial data:', accuracy_score(y_train,
y_pred_adv))
```

In the above code, we first generate a random dataset with 100 samples and 2 features. We then create a SVM classifier with a linear kernel and train it on the original data. We compute the accuracy of the classifier on the original data and print it.

Next, we use the Fast Gradient Sign Method to generate adversarial examples with a perturbation of 0.1. We then train the SVM classifier on the adversarial examples and compute the accuracy on the adversarial data. We can see that the accuracy on adversarial data is lower than that on original data, which indicates that the classifier is more robust to adversarial attacks after adversarial training.

Adversarial training is a powerful technique to improve the robustness of machine learning models against adversarial attacks. By generating and training on adversarial examples, models can learn to recognize and classify such examples correctly and improve their ability to handle adversarial attacks.

8.9 VOTING CLASSIFIER

Ensemble learning methods combine multiple machine learning models to improve the predictive performance of the overall model. One of the simplest and most popular ensemble methods is the Voting Classifier, which combines the predictions of multiple individual classifiers to make a final prediction. In this section, we will discuss the concept of the Voting Classifier and provide a real-life coding example.

What is a Voting Classifier?

A Voting Classifier is an ensemble learning method that combines the predictions of multiple individual classifiers to make a final prediction. The idea behind the Voting Classifier is that by combining the predictions of multiple classifiers, the overall prediction will be more accurate and less prone to errors than any individual classifier.

The Voting Classifier can be implemented in two ways:

Hard voting: In hard voting, each individual classifier makes a binary prediction, and the final prediction is based on the majority vote of the individual predictions.

Soft voting: In soft voting, each individual classifier produces a probability estimate for each class, and the final prediction is based on the average probability of each class across all individual classifiers.

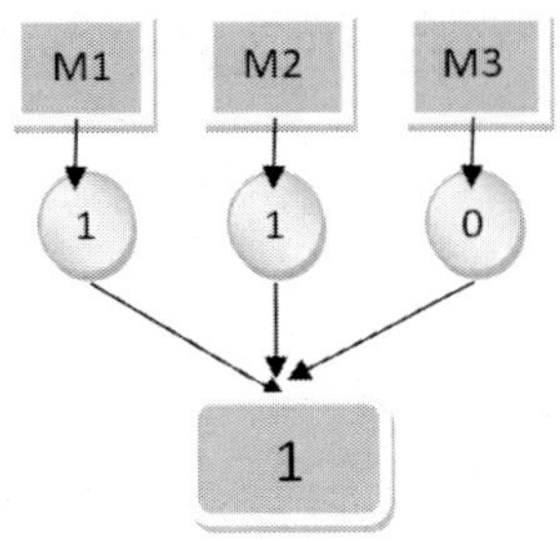

Hard Voting

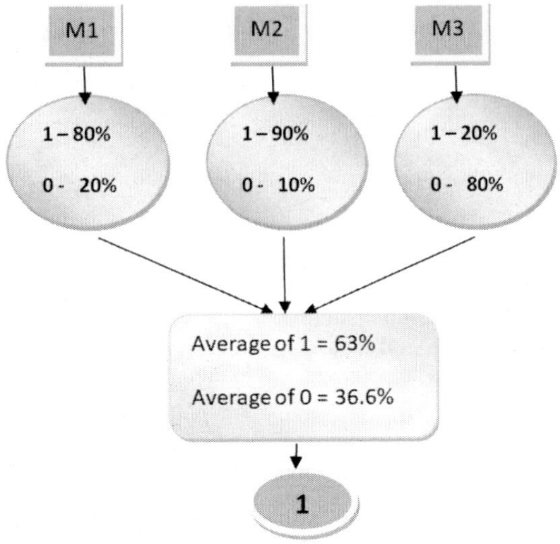

Soft Voting

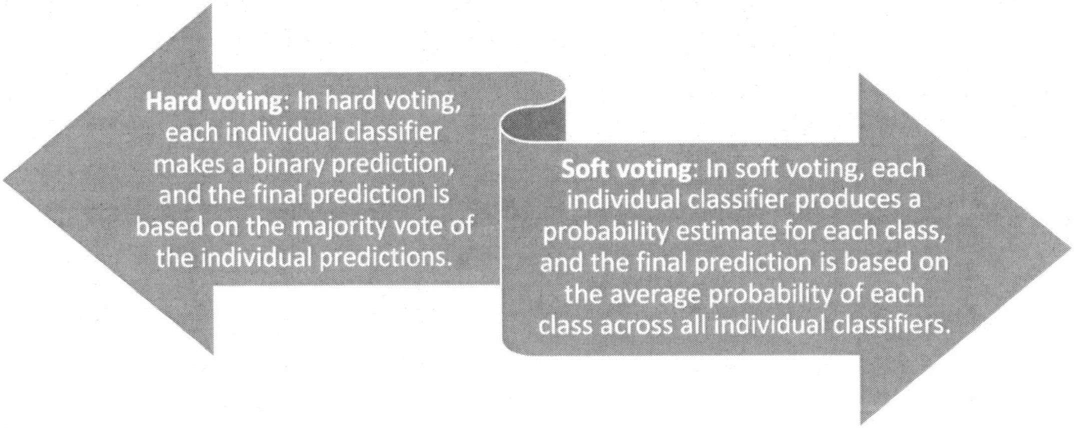

Coding Example:

Let's implement a Voting Classifier on a random dataset using the scikit-learn library. We will first generate a random dataset using the make_classification function of scikit-learn, which generates a random n-class classification problem.

```
from sklearn.datasets import make_classification
from sklearn.model_selection import train_test_split
from sklearn.ensemble import VotingClassifier
from sklearn.tree import DecisionTreeClassifier
from sklearn.linear_model import LogisticRegression
from sklearn.svm import SVC
from sklearn.metrics import accuracy_score

# Generate a random dataset
X, y = make_classification(n_samples=1000, n_classes=2,
random_state=42)

# Split the dataset into training and testing sets
X_train, X_test, y_train, y_test = train_test_split(X, y,
test_size=0.2, random_state=42)

# Define the individual classifiers
clf1 = DecisionTreeClassifier(random_state=42)
clf2 = LogisticRegression(random_state=42)
clf3 = SVC(kernel='linear', probability=True, random_state=42)

# Define the Voting Classifier
voting_clf = VotingClassifier(estimators=[('dt', clf1), ('lr', clf2),
('svm', clf3)], voting='hard')

# Train the Voting Classifier
voting_clf.fit(X_train, y_train)

# Make predictions on the test set
y_pred = voting_clf.predict(X_test)

# Evaluate the accuracy of the Voting Classifier
accuracy = accuracy_score(y_test, y_pred)
print('Accuracy:', accuracy)
```

In the above code, we first generate a random dataset using the make_classification function of scikit-learn. We then split the dataset into training and testing sets using the train_test_split function of scikit-learn. Next, we define three individual classifiers: a Decision Tree Classifier, a Logistic Regression Classifier, and a Support Vector Machine Classifier. We then define the Voting Classifier using the VotingClassifier class of scikit-learn and set the estimators parameter to a list of the individual classifiers. We set the voting parameter to 'hard' to implement hard voting. We then train the Voting Classifier on the training set using the fit method. Finally, we make predictions on the test set using the predict method and evaluate the accuracy of the Voting Classifier using the accuracy_score function of scikit-learn.

8.10 SUMMARY

- Ensemble learning methods combine multiple models to improve overall prediction accuracy and robustness.
- Bagging is a method that uses bootstrapping to create multiple models trained on random subsets of the data.
- Boosting is a method that adapts weak models to the strong model by iteratively training new models on misclassified samples.
- Gradient boosting is a popular boosting method that uses gradient descent optimization to minimize a loss function.
- XGBoost and LightGBM are powerful gradient boosting frameworks that utilize advanced optimization techniques for faster training and improved accuracy.

- Stacking is a method that combines multiple models by training a meta-model on the outputs of the base models.
- Blending is a simpler version of stacking that uses weighted averaging of the base models' predictions.
- Rotation forest is an ensemble method that randomly rotates the feature space and trains multiple models on the transformed data.
- Cascading classifiers is an ensemble method that uses a series of classifiers to classify data, with each subsequent classifier focusing on the misclassified samples of the previous classifier.
- Adversarial training is a method that trains models on adversarial examples, which are purposely crafted inputs designed to fool the model.
- Ensemble learning methods can significantly improve model accuracy and generalization, but can also increase model complexity and training time.

8.11 TEST YOUR KNOWLEDGE

I. Which of the following is an example of an ensemble learning method?

a) Linear Regression
b) Decision Trees
c) Random Forest
d) Naive Bayes

II. Which ensemble method involves combining the predictions of multiple models using a weighted average?

a) Bagging

b) Boosting
c) Stacking
d) Blending

III. What is the primary purpose of ensemble learning methods?

a) To increase the accuracy of a single model
b) To decrease the complexity of a single model
c) To make the model faster to train
d) To reduce overfitting in a model

IV. Which of the following is an example of a meta-learner in a stacking ensemble?

a) Decision Tree
b) Linear Regression
c) Random Forest
d) Gradient Boosting

V. Which ensemble method involves training multiple models on different subsets of the training data?

a) Bagging
b) Boosting
c) Stacking
d) Blending

VI. Which ensemble method involves training new models to correct the errors of previous models?

a) Bagging
b) Boosting
c) Stacking
d) Blending

VII. Which ensemble method involves creating new features by combining the outputs of multiple models?

a) Bagging

b) Boosting
c) Stacking
d) Blending

VIII. Which ensemble method is known for its ability to handle imbalanced datasets?

a) Bagging
b) Boosting
c) Stacking
d) Adversarial Training

IX. Which of the following is a disadvantage of ensemble learning methods?

a) They are computationally expensive
b) They are prone to overfitting
c) They can only be used with certain types of models
d) They are not very accurate

X. Which ensemble method involves training a chain of models, with each model learning to distinguish between the classes that the previous models classified as equal?

a) Bagging
b) Boosting
c) Stacking
d) Cascading Classifiers

8.12 PRACTICAL EXERCISE

A. Build a random forest model on the given dataset and evaluate its performance using cross-validation.

B. Build an AdaBoost model on the given dataset and evaluate its performance using cross-validation.
C. Build an XGBoost model on the given dataset and evaluate its performance using cross-validation.
D. Build a stacking model with a logistic regression meta-estimator on the given dataset and evaluate its performance using cross-validation.
E. Build a voting classifier with a hard voting strategy on the given dataset and evaluate its performance using cross-validation.

F.

8.13 ANSWERS

I. Answer: c) Random Forest
II. Answer: d) Blending
III. Answer: a) To increase the accuracy of a single model
IV. Answer: b) Linear Regression
V. Answer: a) Bagging
VI. Answer: b) Boosting
VII. Answer: c) Stacking
VIII. Answer: b) Boosting
IX. Answer: a) They are computationally expensive
X. Answer: d) Cascading Classifiers

8.14 EXERCISE SOLUTIONS

Solution A:

```
from sklearn.ensemble import RandomForestClassifier
from sklearn.model_selection import cross_val_score
import pandas as pd

# load dataset
data = pd.read_csv('dataset.csv')

# separate features and target
X = data.drop('target', axis=1)
y = data['target']

# initialize random forest classifier
rf_model = RandomForestClassifier()

# evaluate performance using cross-validation
scores = cross_val_score(rf_model, X, y, cv=5)
print("Accuracy scores: ", scores)
print("Mean accuracy score: ", scores.mean())
```

Solution B:

```
from sklearn.ensemble import AdaBoostClassifier
from sklearn.model_selection import cross_val_score
import pandas as pd

# load dataset
data = pd.read_csv('dataset.csv')

# separate features and target
X = data.drop('target', axis=1)
y = data['target']

# initialize AdaBoost classifier
ada_model = AdaBoostClassifier()

# evaluate performance using cross-validation
scores = cross_val_score(ada_model, X, y, cv=5)
print("Accuracy scores: ", scores)
print("Mean accuracy score: ", scores.mean())
```

Solution C:

```
from xgboost import XGBClassifier
from sklearn.model_selection import cross_val_score
import pandas as pd

# load dataset
data = pd.read_csv('dataset.csv')

# separate features and target
X = data.drop('target', axis=1)
y = data['target']

# initialize XGBoost classifier
xgb_model = XGBClassifier()

# evaluate performance using cross-validation
scores = cross_val_score(xgb_model, X, y, cv=5)
print("Accuracy scores: ", scores)
print("Mean accuracy score: ", scores.mean())
```

Solution D:

```
from sklearn.ensemble import RandomForestClassifier,
GradientBoostingClassifier, StackingClassifier
from sklearn.linear_model import LogisticRegression
from sklearn.model_selection import cross_val_score
import pandas as pd

# load dataset
data = pd.read_csv('dataset.csv')

# separate features and target
X = data.drop('target', axis=1)
y = data['target']

# initialize base models
rf_model = RandomForestClassifier()
gb_model = GradientBoostingClassifier()

# initialize stacking model with a logistic regression meta-estimator
stack_model = StackingClassifier(estimators=[('rf', rf_model), ('gb',
gb_model)], final_estimator=LogisticRegression())

# evaluate performance using cross-validation
scores = cross_val_score(stack_model, X, y, cv=5)
print("Accuracy scores: ", scores)
print("Mean accuracy score: ", scores.mean())
```

Solution E:

```
from sklearn.ensemble import RandomForestClassifier,
GradientBoostingClassifier, VotingClassifier
from sklearn.model_selection import cross_val_score
import pandas as pd

# load dataset
data = pd.read_csv('dataset.csv')

# separate features and target
X = data.drop('target', axis=1)
```

```
y = data['target']

# initialize base models
rf_model = RandomForestClassifier()
gb_model = GradientBoostingClassifier()

# initialize voting classifier with a hard voting strategy
vote_model = Voting
```

09

9 REAL-WORLD APPLICATIONS OF MACHINE LEARNING

Machine learning is a rapidly growing field that has the potential to revolutionize various industries. From healthcare to finance, transportation to e-commerce, machine learning is being used to improve decision making, automate processes, and create new products and services. In this chapter, we will explore some real-world applications of machine learning, including the challenges and benefits of implementing these models in various industries. We will also discuss the latest developments and trends in the field, and what to expect from machine learning in the future. By the end of this chapter, you will have a better understanding of the potential and limitations of machine learning and its role in shaping the world around us.

9.1 NATURAL LANGUAGE PROCESSING

Natural Language Processing (NLP) is a subfield of artificial intelligence and computer science that focuses on the interaction between computers and humans using natural language. NLP enables computers to understand, interpret and generate human language, including text, speech and handwriting.

One of the key tasks in NLP is **text classification**, which is the process of assigning predefined categories or labels to a given text. For example, classifying emails as spam or not spam, or news articles as politics or sports.

Another important task in NLP is **named entity recognition (NER)**, which is the process of identifying and classifying named entities in text, such as people, organizations, locations, and so on.

NLP also includes tasks such as **sentiment analysis**, which is the process of determining the emotional tone of a given text, and machine translation, which is the process of automatically translating text from one language to another.

In recent years, deep learning techniques have been applied to NLP and have achieved state-of-the-art results in many NLP tasks, such as **language translation, text summarization, and question answering.**

One popular library for NLP in Python is **NLTK** (Natural Language Toolkit), which provides a wide range of tools and resources for working with human language data. Another popular library is **spaCy**, which is designed specifically for production use and is optimized for performance.

An example of NLP in action is an **email classification** system that uses machine learning algorithms to automatically sort incoming emails into different folders such as "spam" or "not spam". The system would first be trained on a dataset of labeled emails and then be able to classify new incoming emails based on their content and features such as sender and keywords.

In healthcare, NLP can be used to extract useful information from unstructured medical data such as electronic health records, medical notes, and discharge summaries. This can aid in disease diagnosis, treatment planning, and drug discovery. NLP can also be used to monitor social media and extract information about public health concerns, such as tracking the spread of a disease or monitoring vaccine hesitancy.

Overall, NLP is an important application of machine learning that has a wide range of real-world applications in various industries including healthcare, finance, and customer service. With the increasing amount of human language data being generated every day, the need for NLP continues to grow.

A step-by-step use case of NLP

Natural Language Processing (NLP) is a subfield of machine learning that deals with the interaction between computers and human languages. It is an application of machine learning that is used to extract insights from unstructured data in the form of text, speech or any other form of natural language. NLP is used in a wide range of applications such as sentiment analysis, text classification, language translation, and more.

A common use case of NLP is sentiment analysis. Sentiment analysis is the process of determining the emotional tone of text. This can be used in a variety of applications such as social media monitoring, brand management, and customer service. In this use case, we will go through the steps to build a sentiment analysis model that can classify text as positive, negative, or neutral.

Step 1: Collect and Preprocess the Data

The first step is to collect the data. The data should be in the form of text, such as tweets, reviews, or any other form of text data. The data should be labeled with the sentiment (positive, negative or neutral) so that it can be used for training the model. Once the data is collected, it needs to be preprocessed. This includes cleaning the data, removing any special characters, stop words and stemming the words.

Step 2: Vectorize the Text Data

The next step is to vectorize the text data. Vectorization is the process of converting text into numerical form so that it can be used for training the model. There are multiple ways to vectorize text data such as Count Vectorization, Tf-idf Vectorization, and Word Embeddings. Count Vectorization and Tf-idf Vectorization are based on the frequency of words in the text. Word Embeddings are more advanced techniques that consider the context of the words.

Step 3: Split the Data into Training and Testing Sets

Once the data is vectorized, it needs to be split into training and testing sets. The training set will be used to train the model, while the testing set will be used to evaluate the model. This step is important to avoid overfitting of the model.

Step 4: Train the Model

The next step is to train the model. A common model used for sentiment analysis is the logistic regression model. Other models such as Random Forest, Support Vector Machines, and Naive Bayes can also be used. The model should be trained on the training set and then evaluated on the testing set.

Step 5: Evaluate the Model

Once the model is trained, it needs to be evaluated. The evaluation should be done on the testing set. Common metrics used for evaluation include accuracy, precision, recall, and F1-score.

Step 6: Fine-Tune the Model

If the model's performance is not satisfactory, it needs to be fine-tuned. This can be done by adjusting the model's parameters, changing the vectorization method or collecting more data.

Step 7: Deploy the Model

Once the model's performance is satisfactory, it can be deployed to be used in a real-world application. The model can be integrated into a web or mobile application to classify text as positive, negative, or neutral.

In this use case, we have gone through the steps to build a sentiment analysis model using NLP. Sentiment analysis is just one application of NLP, there are many other applications such as language translation, text summarization, and more. The process of building a model for these applications is similar, but the data and the model used will be different.

9.2 COMPUTER VISION

Computer vision is a subfield of artificial intelligence that deals with the development of algorithms and models that can interpret, understand, and analyze visual data from the world around us. This includes images and videos, and can also include other forms of data such as depth maps, lidar data, and thermal imaging.

One of the key challenges in computer vision is to develop models that can understand and interpret visual data in a way that is similar to how humans do it. This requires the development of models that can recognize objects, identify patterns, and make predictions based on visual data.

There are many different techniques and algorithms that are used in computer vision, including image processing, feature extraction, object recognition, and deep learning. Some of the most popular deep learning architectures for computer vision include convolutional neural networks (CNNs) and recurrent neural networks (RNNs).

In recent years, computer vision has been successfully applied in many different areas, such as:

- **Image classification**: recognizing objects and scenes in images.
- **Object detection**: detecting and locating objects in images and videos.
- **Semantic segmentation**: assigning a semantic label to each pixel in an image.
- **Video analysis**: analyzing and understanding videos and video streams.
- **Autonomous systems**: guiding self-driving cars, drones and robots.
- **Augmented reality**: overlaying digital information onto the real world.

One example use case of computer vision is in retail business. A retail store can use computer vision to track customers in the store, monitor which products they are interacting with, and gather data on how long they spend in each area. This data can then be used to optimize the store layout, improve product placement, and create targeted marketing campaigns.

One of the most popular computer vision application is image classification. For example, a model can be trained to recognize objects like cars, buildings, and animals in images, and then used to automatically label new images with the same objects.

To start working on a computer vision project, it is important to have a clear understanding of the problem you are trying to solve and the data you have available. This will help you to choose the appropriate techniques and algorithms for your project.

Next, you will need to preprocess and prepare your data for training. This may include tasks such as resizing images, normalizing pixel values, and splitting your data into training and testing sets.

Once your data is ready, you can begin training your model using various techniques and algorithms. It is important to evaluate your model using metrics such as accuracy, precision, and recall to ensure that it is performing well.

Finally, you can deploy your model in a production environment and begin using it to make predictions on new data. It is important to monitor the performance of your model in the production environment and make adjustments as needed. Overall, creating a computer vision project requires a lot of experimentation, iteration and fine-tuning to get the best results. With the right approach, it can be a challenging but rewarding experience that can lead to the development of powerful and impactful applications.

To sum up, computer vision is a rapidly growing field of machine learning, with many exciting applications and a lot of opportunities for innovation. With the right skills and resources, anyone can start working on a computer vision project and make a real impact in the world.

9.3 RECOMMENDER SYSTEMS

Recommender systems are a type of machine learning application that are designed to predict the preferences or ratings that a user would give to a particular item. They are commonly used in a variety of applications such as e-commerce websites, music and video streaming services, and social media platforms.

There are several techniques used to build recommender systems, including collaborative filtering, content-based filtering, and hybrid approaches that combine both techniques.

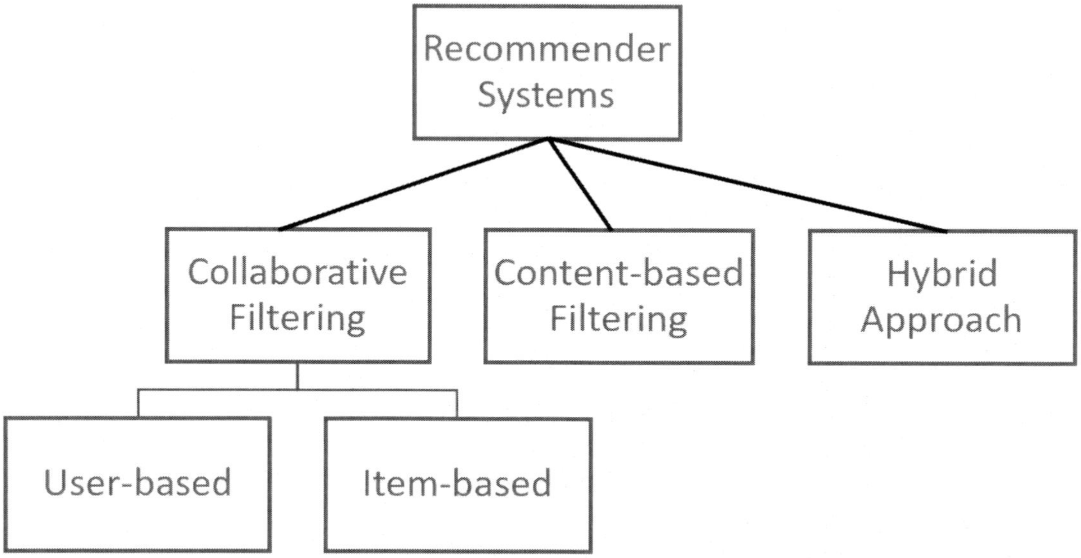

Collaborative filtering is based on the idea that users who have similar preferences in the past will have similar preferences in the future. This technique can be further divided into two sub-techniques: user-based and item-based. In user-based collaborative filtering, the system finds the set of users who are most similar to the active user and recommends items that those users have liked. In item-based collaborative filtering, the system finds the set of items that are most similar to the items the active user has liked, and recommends those items.

Content-based filtering is based on the idea that users will prefer items that are similar to items they have liked in the past. This technique uses the characteristics or attributes of the items to recommend similar items to the user.

Hybrid approaches combine both collaborative filtering and content-based filtering techniques to make recommendations. These approaches can be more effective than either technique alone, as they can take into account both the user's preferences and the characteristics of the items.

To start a machine learning project for a recommender system, one can begin by gathering and preprocessing the data. This includes collecting the ratings or preferences of users for a particular item, as well as any other relevant information such as demographic data about the users or characteristics of the items. Next, one can select and implement a suitable algorithm for the recommender system, such as collaborative filtering or a hybrid approach. Finally, one can evaluate the performance of the recommender system using metrics such as precision and recall, and make any necessary adjustments to improve its performance.

In practice, recommender systems can be quite complex and may involve large amounts of data and computational resources. There are also many variations and modifications that can be made to the basic algorithms to improve performance and address specific challenges, such as dealing with sparse data or handling the scalability of the system. Furthermore, the accuracy of recommender systems can be affected by biases and other factors, such as data availability and user engagement, that need to be considered when designing and implementing the system.

Recommender systems are widely used in various applications such as e-commerce, streaming services, and social media platforms. Some examples of how recommender systems are used in daily life are:

1. **Online Shopping**: Recommender systems are used to suggest products to customers based on their browsing and purchase history. For example,

Amazon suggests products to customers based on their purchase history and the products they have viewed.

2. **Streaming Services**: Recommender systems are used to suggest movies, TV shows and music to users based on their watch and listen history. For example, Netflix suggests TV shows and movies to users based on their watch history and the genres they have shown interest in.

3. **Social Media**: Recommender systems are used to suggest friends and pages to users based on their browsing and friending history. For example, Facebook suggests friends and pages to users based on their friending history and the pages they have liked.

To implement a recommender system, one can follow these steps:

1. **Gather data**: Collect data on user preferences, browsing history, and purchase history. This can be done by tracking user interactions on your website or app.

2. **Preprocess data**: Clean and preprocess the data to remove missing values, duplicate records, and outliers.

3. **Exploratory Data Analysis (EDA)**: Perform EDA on the data to understand the patterns and relationships in the data.

4. **Select a model**: Select an appropriate model based on the type of data and the problem you are trying to solve. Some popular models include collaborative filtering, content-based filtering, and hybrid models.

5. **Train and test the model**: Train the model on the collected data and test it on a separate dataset.

6. **Deploy the model**: Deploy the model in production and track its performance to fine-tune it as necessary.

9.4 TIME SERIES FORECASTING

Time series forecasting is a technique used to predict future values based on previously observed values. It is commonly used in various fields such as finance, economics, weather forecasting, and more. Time series forecasting can be approached using various machine learning algorithms, including linear regression, ARIMA, and neural networks.

One common use case of time series forecasting is in the field of finance. For example, stock market forecasting is an important task for investors and traders. They can use historical stock prices and other financial data to predict the future prices of stocks. By analyzing trends and patterns in the data, a machine learning model can make predictions about future stock prices.

Another use case of time series forecasting is in weather forecasting. Meteorologists can use historical weather data, such as temperature, precipitation, and wind speed, to predict future weather patterns. By analyzing patterns in the data, a machine learning model can make predictions about future weather conditions.

To implement a time series forecasting project, one can follow the following steps:

1. **Collect and clean the data:** Gather historical time series data relevant to the problem at hand. Clean and preprocess the data to remove any missing values, outliers, or irrelevant information.

2. **Explore the data**: Use visualization techniques to explore the data and understand the underlying patterns and trends. Identify any seasonality or trend in the data.

3. **Select a model**: Choose a suitable machine learning model based on the data and the problem. Some popular models for time series forecasting include linear regression, ARIMA, and neural networks.

4. **Train the model**: Train the selected model on the historical data.

5. **Make predictions:** Use the trained model to make predictions about future values.

6. **Evaluate the model**: Evaluate the performance of the model using metrics such as mean absolute error, mean squared error, or root mean squared error.

7. **Fine-tune the model**: Based on the evaluation results, fine-tune the model to improve its performance.

8. **Deploy the model**: Once the model is fine-tuned, it can be deployed in a production environment to make predictions in real-time.

It's worth noting that time series forecasting is a complex task and requires a good understanding of the underlying data and the problem, as well as a lot of experimentation. In addition, it's important to use appropriate techniques for handling seasonality and trend in the data, for example, by using techniques like differencing, decomposition, and so on.

9.5 PREDICTIVE MAINTENANCE

Predictive maintenance is a powerful application of machine learning that enables organizations to predict when equipment or systems are likely to fail, so they can be repaired or replaced before they cause major disruptions or downtime. By analyzing data from various sources, such as sensor readings, machine logs, and historical maintenance records, predictive maintenance algorithms can detect patterns and anomalies that indicate potential problems.

One of the key benefits of predictive maintenance is that it can help organizations avoid the high costs associated with unexpected downtime. For example, in a manufacturing setting, a machine breakdown can lead to lost production, delayed deliveries, and increased labor costs. Similarly, in a transportation or logistics setting, a vehicle breakdown can lead to delays and additional costs for repairs or replacement.

The process of implementing predictive maintenance typically involves several steps:

1. **Data collection**: The first step is to gather data from the equipment or systems that will be monitored. This data can include sensor readings, machine logs, and historical maintenance records.

2. **Data cleaning**: Once the data is collected, it needs to be cleaned and preprocessed to ensure that it is in a format that can be used by the machine learning algorithms. This may involve removing missing or duplicate data, handling outliers, or transforming the data into a more suitable format.

3. **Feature engineering**: The next step is to extract relevant features from the data that will be used to train the machine learning models. This may involve creating new features by combining existing ones, or selecting a subset of the data that is most relevant to the problem.

4. **Model selection and training**: Once the data is prepared, the next step is to select the appropriate machine learning model and train it on the data. This may involve trying different models and evaluating their performance using different evaluation metrics.

5. **Deployment and monitoring**: After the model is trained, it needs to be deployed in a production environment and monitored for performance. This may involve setting up automated alerts to notify the team of potential issues, or creating dashboards to visualize the model's performance.

6. **Continual improvement**: The final step is to continually monitor and improve the model over time by updating it with new data and retraining as needed.

Overall, predictive maintenance is a powerful application of machine learning that can help organizations improve their equipment and systems' reliability and avoid unexpected downtime. By using machine learning models to analyze data from various sources, organizations can detect patterns and anomalies that indicate potential problems, and take preventative action before they cause major disruptions.

9.6 SPEECH RECOGNITION

Speech recognition, also known as automatic speech recognition or ASR, is a form of artificial intelligence that allows machines to recognize and transcribe spoken language. It is used in a wide range of applications, such as virtual assistants, voice-controlled devices, and call centers.

Deep learning techniques, particularly recurrent neural networks (RNNs) and long short-term memory (LSTM) networks, have been widely adopted in speech recognition systems. These models are able to handle the complex variations in speech patterns and are able to learn from large amounts of data.

One of the key challenges in speech recognition is the variability in speech patterns due to different accents, speaking styles, and background noise. To overcome this, many systems use a combination of techniques, such as acoustic modeling, language modeling, and phoneme recognition.

Acoustic modeling involves training a model on a large dataset of audio samples to learn the underlying patterns of speech. Language modeling involves training a model to predict the likelihood of a sequence of words, given the context. Phoneme recognition involves breaking down speech into its individual sounds, known as phonemes, and then recognizing them.

There are various open-source libraries and frameworks available for speech recognition, such as Kaldi, CMUSphinx, and HTK. Additionally, there are many cloud-based speech recognition services available, such as Google Cloud Speech-to-Text, Amazon Transcribe, and Microsoft Azure Speech Services.

Speech recognition has a wide range of applications in various industries, including healthcare, automotive, and customer service. In healthcare, speech recognition can be used to transcribe medical dictations, allowing for more efficient documentation and record-keeping. In the automotive industry, speech recognition can be used in cars to control various functions, such as navigation and entertainment. In customer service, speech recognition can be used to improve call center operations by allowing customers to interact with a computer-based agent.

9.7 ROBOTICS AND AUTOMATION

Robotics and Automation is a field that has seen a significant impact from machine learning. Machine learning techniques such as reinforcement learning and deep learning have been used to train robots to perform tasks such as grasping, manipulation, and navigation.

One of the most significant applications of machine learning in robotics is in **industrial automation**. Industrial robots are used in a wide range of industries such as manufacturing, logistics, and agriculture to perform repetitive and dangerous tasks. Machine learning techniques have been used to improve the performance of these robots by allowing them to adapt to changes in their environment and learn new tasks.

In addition, machine learning has also been used to develop autonomous systems such as self-driving cars and drones. These systems use a variety of sensors such as cameras, lidar, and radar to perceive their environment and make decisions. Machine learning algorithms are used to process sensor data and make predictions about the environment, such as the location of other vehicles or obstacles.

Another application of machine learning in robotics is in human-robot interaction. Machine learning algorithms are used to process data from sensors such as microphones and cameras to recognize human speech and gestures. This allows robots to understand and respond to human commands, making them more user-friendly and accessible.

In summary, machine learning has played a crucial role in the field of robotics and automation by enabling robots to adapt to changing environments, learn new tasks, and interact with humans more effectively. As the field of machine learning continues to evolve, we can expect to see even more advanced applications of robotics and automation in various industries.

9.8 AUTONOMOUS DRIVING

Autonomous driving is a rapidly growing field that utilizes machine learning techniques to enable vehicles to drive themselves without human intervention. Some key components of autonomous driving include object detection, path planning, and control.

Object detection is the process of identifying and locating objects, such as other vehicles, pedestrians, and road signs, in images or video. This information is used to understand the vehicle's surroundings and make decisions about how to navigate the environment.

Path planning is the process of determining the best path for the vehicle to take based on its current location, the location of detected objects, and the vehicle's desired destination. This step involves decision-making, such as determining when to slow down, speed up, change lanes, and make turns.

Control is the process of executing the path planned by the vehicle. This includes sending commands to the vehicle's actuators, such as the steering, throttle, and brakes, to ensure the vehicle follows the planned path.

Autonomous vehicles are already being tested on public roads, and it is expected that they will be widely available in the near future. These vehicles have the potential to significantly improve road safety and reduce traffic congestion. However, it is important that they are thoroughly tested and proven to be safe before they are released to the public.

9.9 FRAUD DETECTION

Fraud detection is an important application of machine learning in various industries such as finance, e-commerce, and insurance. Fraudulent activities can cause significant financial loss and damage to a company's reputation. Traditional methods of fraud detection, such as manual reviews and rule-based systems, can be time-consuming and may not be able to detect all types of fraud. Machine learning algorithms, on the other hand, can analyze large amounts of data and detect patterns and anomalies that may indicate fraud.

Some common machine learning techniques used in fraud detection include:

- Anomaly detection: This technique can identify transactions or patterns that deviate from the normal behavior. For example, a transaction with a large amount or an unusual location may be flagged as potentially fraudulent.
- Clustering: This technique can group similar transactions together and identify clusters of suspicious activity.
- Supervised learning: This technique can learn from labeled data, such as past fraudulent transactions, to classify new transactions as fraudulent or not.
- Deep learning: This technique can analyze large amounts of data, such as images or text, to detect fraud.

One example of the application of machine learning in fraud detection is the use of credit card fraud detection. Here, machine learning algorithms are trained on historical transaction data to identify patterns and anomalies that indicate fraud. The model is then deployed to monitor and detect any suspicious transactions in real-time.

It is worth noting that as fraudsters are always looking for new ways to perpetrate fraud, machine learning models used for fraud detection must be continuously updated and retrained to stay current with the latest fraudulent tactics.

9.10 OTHER REAL-LIFE APPLICATIONS

- **Anomaly Detection:** Machine learning can be used to detect anomalies in data, such as unexpected spikes or drops in a time series, or unusual patterns in images or videos. These anomalies can indicate potential problems or opportunities that require further investigation.

- **Image and Video Analysis**: Machine learning algorithms can be used to analyze images and videos, such as recognizing objects and faces, detecting patterns, and tracking movements. This can be applied in areas such as surveillance, medical imaging, and self-driving cars.

- **Text Classification**: Machine learning algorithms can be used to classify and categorize text, such as emails, documents, and social media posts. This can be applied in areas such as sentiment analysis, spam detection, and topic modeling.

- **Prediction in Finance**: Machine learning can be used to predict stock prices, currency exchange rates, and other financial variables. This can be applied in areas such as risk management and portfolio optimization.

- **Weather Forecasting**: Machine learning can be used to predict weather patterns and forecast conditions such as temperature, precipitation, and wind. This can be applied in areas such as agriculture, construction, and emergency management.

- **Customer Segmentation**: Machine learning can be used to segment customers into groups based on their demographics, behavior, and preferences. This can be applied in areas such as marketing, sales, and customer service.

- **Marketing Personalization**: Machine learning can be used to personalize marketing messages and offers, based on customer data and behavior. This can be applied in areas such as email marketing, social media advertising, and e-commerce.
- **Supply Chain Optimization**: Machine learning can be used to optimize and streamline supply chain operations by predicting demand, identifying bottlenecks, and improving inventory management.
- **Manufacturing Optimization**: Machine learning can be used to optimize manufacturing processes by predicting equipment failures, reducing downtime, and improving overall efficiency.
- **Predictive Maintenance**: Machine learning can be used to predict when equipment or machinery is likely to fail, allowing for proactive maintenance and reducing downtime.
- **Energy Forecasting**: Machine learning can be used to predict energy demand and optimize energy usage, helping to reduce costs and improve sustainability.
- **Agriculture Optimization**: Machine learning can be used to optimize crop yields, predict weather patterns, and improve overall efficiency in the agriculture industry.
- **Environmental Monitoring**: Machine learning can be used to analyze data from sensors and other monitoring devices to predict and prevent environmental hazards, such as air and water pollution.
- **Crime Prediction**: Machine learning algorithms can be used to analyze historical crime data to predict where and when crimes are likely to occur

in the future. This can help law enforcement agencies to better allocate resources and reduce crime rates.

- **Traffic Flow Prediction**: Machine learning can be used to analyze traffic data and predict traffic flow patterns in real-time. This can help traffic authorities to optimize traffic signal timings, reduce congestion, and improve overall traffic flow.
- **Sports Analytics**: Machine learning can be used to analyze sports data and gain insights into player performance, team strategies, and game outcomes. This can help coaches and managers to make better decisions and improve team performance.
- **Gaming Analytics**: Machine learning can be used to analyze player data and gain insights into player behavior and preferences. This can help game developers to optimize game design and improve player engagement.
- **Social Media Analysis**: Machine learning can be used to analyze social media data and gain insights into consumer behavior and preferences. This can help businesses to better understand and target their audience, improve marketing campaigns, and increase sales.
- **Cybersecurity**: Machine learning can be used to analyze network data and detect potential security threats. This can help organizations to protect against cyber attacks and improve overall security.
- **E-commerce**: Machine learning can be used to analyze customer data and gain insights into consumer behavior and preferences. This can help e-commerce businesses to better understand and target their audience, improve marketing campaigns, and increase sales.

9.11 SUMMARY

- Machine learning is increasingly being applied in healthcare to improve patient outcomes and streamline healthcare operations.
- Natural Language Processing (NLP) is a popular application of machine learning in areas such as language translation, sentiment analysis, and text summarization.
- Computer vision uses machine learning techniques to analyze and understand images and videos, with applications in areas such as self-driving cars, surveillance, and image search.
- Recommender systems are widely used in e-commerce and media, to personalize user experience and recommend products or content.
- Time series forecasting and Predictive maintenance are used in industries such as finance, manufacturing, and energy to predict future events and optimize operations.
- Robotics and Automation is another area where machine learning is being applied to improve efficiency and productivity.
- Machine learning is also being applied in other fields such as Fraud Detection, Anomaly Detection, Image and Video Analysis, Speech Recognition, Text Classification, Prediction in finance, Autonomous Driving, Weather forecasting, Customer Segmentation and Marketing personalization.
- Other applications of machine learning include Supply Chain Optimization, Manufacturing Optimization, Predictive Maintenance,

Energy Forecasting, Agriculture Optimization and Environmental Monitoring.

- Machine learning is also being used in crime prediction, traffic flow prediction, sports analytics, gaming analytics, social media analysis, cybersecurity, and e-commerce.

9.12 TEST YOUR KNOWLEDGE

I. **Which of the following is not an application of machine learning in healthcare?**
  a) Identifying potential outbreaks of infectious diseases
  b) Performing surgery
  c) Analyzing medical images
  d) Monitoring patient vital signs

II. **Which industry commonly uses recommender systems?**
  a) Healthcare
  b) Retail
  c) Agriculture
  d) Manufacturing

III. **Which of the following is not a common application of machine learning in robotics and automation?**
  a) Autonomous driving
  b) Industrial automation
  c) Weather forecasting
  d) Predictive maintenance

IV. **Which of the following is not an application of machine learning in finance?**
  a) Risk assessment

b) Fraud detection
c) Stock market prediction
d) Budget forecasting

V. Which of the following is not an application of machine learning in agriculture?

a) Crop yield prediction
b) Livestock monitoring
c) Autonomous tractors
d) Weather forecasting

VI. Which of the following is not an application of machine learning in supply chain optimization?

a) Inventory management
b) Delivery route optimization
c) Warehouse layout design
d) Speech recognition

VII. Which of the following is not an application of machine learning in cybersecurity?

a) Intrusion detection
b) Network monitoring
c) Spam filtering
d) Game development

VIII. Which of the following is not an application of machine learning in e-commerce?

a) Product recommendation
b) Fraud detection
c) Inventory management
d) Autonomous driving

IX. Which of the following is not an application of machine learning in sports analytics?

a) Player performance analysis
b) Game strategy optimization
c) Weather forecasting
d) Fan engagement

X. **Which of the following is not an application of machine learning in social media analysis?**

a) Sentiment analysis
b) Content recommendation
c) Ad targeting
d) 3D modeling

XI. **Which field commonly uses machine learning for anomaly detection?**

a) Healthcare
b) Finance
c) Agriculture
d) Cybersecurity

XII. **What is one application of machine learning in the manufacturing industry?**

a) Supply chain optimization
b) Fraud detection
c) Speech recognition
d) Predictive maintenance

XIII. **Which industry commonly uses machine learning for recommender systems?**

a) Healthcare
b) E-commerce
c) Sports analytics
d) Agriculture

XIV. **What is one application of machine learning in the energy industry?**

a) Autonomous driving
b) Crime prediction
c) Energy forecasting
d) Social media analysis

XV. **Which field commonly uses machine learning for image and video analysis?**

a) Robotics and automation
b) Cybersecurity
c) Agriculture
d) Surveillance

XVI. **What is one application of machine learning in the transportation industry?**

a) Fraud detection
b) Traffic flow prediction
c) Speech recognition
d) Gaming analytics

XVII. **Which industry commonly uses machine learning for predictive maintenance?**

a) Healthcare
b) E-commerce
c) Manufacturing
d) Agriculture

XVIII. **What is one application of machine learning in the environment industry?**

a) Autonomous driving
b) Crime prediction
c) Environmental monitoring

d) Social media analysis

XIX. **Which field commonly uses machine learning for text classification?**

a) Healthcare
b) Finance
c) Agriculture
d) Cybersecurity

XX. **What is one application of machine learning in the agriculture industry?**

a) Supply chain optimization
b) Fraud detection
c) Speech recognition
d) Agriculture optimization

XXI. **Which industry commonly uses machine learning for speech recognition?**

a) Healthcare
b) E-commerce
c) Automotive
d) Agriculture

XXII. **What is one application of machine learning in finance industry?**

a) Predictive Maintenance
b) Fraud Detection
c) Predictive Analysis
d) Gaming analytics

XXIII. **Which field commonly uses machine learning for customer segmentation?**

a) E-commerce
b) finance

c) Manufacturing
d) Healthcare

XXIV. What is one application of machine learning in the marketing industry?

a) Supply chain optimization
b) Fraud detection
c) Speech recognition
d) Marketing Personalization

XXV. Which industry commonly uses machine learning for time series forecasting?

a) Healthcare
b) Finance
c) Energy
d) Sports analytics

XXVI. What is one application of machine learning in the gaming industry?

a) Gaming Analytics
b) Traffic flow prediction
c) Environmental monitoring
d) Social media analysis

9.13 ANSWERS

I.	Answer: b)	Performing surgery
II.	Answer: b)	Retail
III.	Answer: c)	Weather forecasting
IV.	Answer: d)	Budget forecasting
V.	Answer: d)	Weather forecasting
VI.	Answer: d)	Speech recognition
VII.	Answer: d)	Game development
VIII.	Answer: d)	Autonomous driving
IX.	Answer: c)	Weather forecasting
X.	Answer: d)	3D modeling
XI.	Answer: b)	Finance
XII.	Answer: d)	Predictive maintenance
XIII.	Answer: b)	E-commerce
XIV.	Answer: c)	Energy forecasting
XV.	Answer: d)	Surveillance
XVI.	Answer: b)	Traffic flow prediction
XVII.	Answer: c)	Manufacturing
XVIII.	Answer: c)	Environmental monitoring
XIX.	Answer: d)	Cybersecurity
XX.	Answer: d)	Agriculture optimization
XXI.	Answer: c)	Automotive
XXII.	Answer: c)	Predictive Analysis
XXIII.	Answer: a)	E-commerce
XXIV.	Answer: d)	Marketing Personalization
XXV.	Answer: b)	Finance
XXVI.	Answer: a)	Gaming Analytics

Appendix A

A. FUTURE DIRECTIONS IN PYTHON MACHINE LEARNING

The field of machine learning is rapidly evolving, and Python has become one of the most popular programming languages for developing machine learning models. In recent years, there have been several advancements and new trends in Python machine learning that are worth mentioning.

1. **Deep Learning Frameworks**: There are several deep learning frameworks available in Python such as TensorFlow, Keras, PyTorch, and Caffe, which have made it easier to build complex neural network models. These frameworks have simplified the process of building, training, and deploying deep learning models.

2. **AutoML**: Automated Machine Learning (AutoML) is a rapidly growing trend in the field of machine learning. AutoML tools automate the process of selecting the best algorithm and hyperparameter tuning, which saves time and effort for data scientists.

3. **Explainable AI**: With the increasing use of machine learning models in critical decision-making, there is a growing need for explainable AI. This trend is focused on developing machine learning models that can provide clear explanations for their predictions.

4. **Reinforcement Learning**: Reinforcement learning is a type of machine learning that focuses on training models to make decisions through trial and error. This technique is being applied to a wide range of applications, including robotics, gaming, and finance.

5. **Generative Models**: Generative models, such as Generative Adversarial Networks (GANs) and Variational Autoencoders (VAEs) are becoming increasingly popular in the field of machine learning. These models can generate new data samples that are similar to the training data, which can be used for a variety of applications such as image synthesis, text generation, and more.

6. **Transfer Learning**: Transfer learning is a technique that enables models trained on one task to be applied to another, similar task. This technique has been used to improve the performance of natural language processing, computer vision, and speech recognition models.

7. **Edge Computing**: With the increasing use of machine learning in IoT devices and edge computing, there is a growing need for machine learning models that can be deployed on edge devices. This trend is focused on developing machine learning models that can run on low-power devices with limited computational resources.

8. **Distributed Training**: With the increasing size of datasets and complexity of models, there is a growing need for distributed training. This technique allows for the parallelization of the training process across multiple devices, which can significantly reduce the training time for large models.

In conclusion, the field of Python machine learning is rapidly evolving and there are many new trends and advancements that are worth keeping an eye on. These trends are making it easier to build, train, and deploy machine learning models, and they are also opening up new possibilities for machine learning applications in various domains.

Appendix B

B. ADDITIONAL RESOURCES

The field of machine learning and artificial intelligence is constantly evolving, and it can be difficult to keep up with the latest developments and best practices. In this Appendix, we will provide a list of additional resources that can help you to continue your learning journey and stay up-to-date with the latest trends and techniques in the field. These resources include books, tutorials, courses, blogs, and other materials that cover a wide range of topics related to machine learning and AI. Whether you are a beginner or an experienced practitioner, these resources will provide valuable insights and information to help you improve your skills and advance your career.

WEBSITES & BLOGS

This section includes a list of websites and blogs that provide valuable information and tutorials on Python machine learning.

Websites and blogs are a great way to stay updated on the latest developments and techniques in machine learning. They also offer a wealth of tutorials, examples, and code snippets that can help you understand the concepts better and apply them to your own projects. Some of the most popular websites and blogs in the field of Python machine learning include:

1. Machine Learning Mastery (https://machinelearningmastery.com/) - A website that provides tutorials, courses, and practical tips for machine learning practitioners.

2. KDnuggets (https://www.kdnuggets.com/) - A website that offers various tutorials and articles on machine learning, data science, and artificial intelligence.

3. DataCamp (https://www.datacamp.com/) - A website that offers online courses and tutorials on data science and machine learning.

4. Data Science Central (https://www.datasciencecentral.com/) - A website that provides articles, tutorials, and resources for data scientists and machine learning practitioners.

5. DataScience.com (https://www.datascience.com/) - A website that provides articles, tutorials, and resources for data scientists and machine learning practitioners.

6. PyData (https://pydata.org/) - PyData is a community-driven platform that offers tutorials, resources, and events on data science and machine learning in Python.

7. Analytics Vidhya (https://www.analyticsvidhya.com/) - Analytics Vidhya is a platform that provides tutorials, articles, and resources on data science and machine learning in Python.

These are just a few examples of the many websites and blogs that can help you continue your learning journey in Python machine learning. Be sure to check them out and explore other resources that can help you improve your skills and knowledge in this field.

ONLINE COURSES AND TUTORIALS

Online courses and tutorials are a great way to learn about machine learning and Python. There are a wide variety of resources available, from free tutorials to paid courses, and from beginner to advanced levels. Some popular options include:

- **Coursera**: This website offers a wide variety of online courses on machine learning and related topics, taught by professors from top universities. Some popular options include "Machine Learning" by Andrew Ng and "Introduction to Machine Learning" by Katie Bouman.
- **edX**: This website offers a wide variety of online courses on machine learning and related topics, taught by professors from top universities. Some popular options include "Introduction to Machine Learning" by Yaser Abu-Mostafa and "Deep Learning Fundamentals" by IBM.
- **DataCamp**: This website offers a wide variety of online courses and tutorials on machine learning and related topics, taught by experts in the field. Some popular options include "Introduction to Machine Learning" and "Deep Learning in Python".
- **YouTube**: There are many YouTube channels that offer tutorials and lectures on machine learning and Python, such as "Sentdex" and "Codebasics".
- **Udemy**: This website offers a wide variety of online courses on machine learning and related topics, taught by experts in the field. Some popular options include "Python for Data Science and Machine Learning Bootcamp" and "Deep Learning A-Z: Hands-On Artificial Neural Networks"

- **Kaggle**: Kaggle is a platform for data science competitions, with a large community of data scientists sharing tutorials, kernels and best practices. They offer many free and paid tutorials on machine learning and related topics, such as "Deep Learning with Python" and "Introduction to Machine Learning".

Overall, there are many online resources available for learning about machine learning and Python, so it's important to find the one that works best for you.

CONFERENCES AND MEETUPS

Conferences and meetups are great opportunities for machine learning practitioners and enthusiasts to come together and share their knowledge, experiences, and ideas. These events provide a platform for attendees to learn about the latest developments in the field, network with like-minded individuals, and gain insights from industry experts.

Some popular machine learning conferences include:

- NeurIPS (Neural Information Processing Systems)
- ICML (International Conference on Machine Learning)
- ICLR (International Conference on Learning Representations)
- CVPR (Computer Vision and Pattern Recognition)
- ACL (Association for Computational Linguistics)

In addition to these large-scale conferences, there are also many smaller, local meetups and workshops that take place in different cities around the world. These events are a great way to connect with others in the machine learning community, learn about new techniques and tools, and stay up-to-date on the latest developments in the field.

Attending these conferences and meetups can be a great way to stay informed about the latest developments in the field, network with other professionals, and gain insights from industry experts. They are a great opportunity for anyone interested in machine learning to learn about the latest research, techniques and tools, and to network with others in the field.

COMMUNITIES AND SUPPORT GROUPS

Communities and support groups are great resources for machine learning practitioners and enthusiasts to connect, learn, and collaborate with others in the field. These groups can be found both online and offline, and they range from local meetups to international conferences.

Online communities and support groups can be found on platforms such as GitHub, Reddit, and Stack Overflow. These groups provide a place for users to ask questions, share resources and knowledge, and collaborate on projects. They also provide a forum for users to connect with others who share similar interests and expertise.

Offline communities and support groups often take the form of meetups and conferences. Meetups are local gatherings where people can come together to learn, network, and share ideas. They are often organized by volunteers and can be found in most major cities around the world. Conferences are larger gatherings that bring together experts and practitioners from around the world to share their knowledge and insights.

Communities and support groups are a great way to stay up-to-date with the latest trends and developments in machine learning, as well as to connect with others who share your interests. They provide a platform for learning, collaboration, and support that can help you to advance your skills and knowledge in the field. Some popular communities and support groups include:

- Kaggle
- Data Science Central
- Data Science Society
- Data Science Society (DSS)
- Data Science Community (DSC)
- Data Science Society (DSS)

These groups can be found on different platforms like LinkedIn, Facebook, Meetup and many more.

PODCASTS

Podcasts can be a great resource for learning about machine learning and staying up to date on the latest developments in the field. Some popular podcasts that cover machine learning and related topics include:

- **Data Skeptic**: A podcast that explores the ways in which data science and machine learning are changing the world, with a focus on the practical applications of these technologies.

- **Machine Learning Guide**: A podcast that provides an introduction to machine learning, with episodes that cover the basics of supervised and unsupervised learning, deep learning, and more.
- **Linear Digressions**: A podcast that covers data science and machine learning, with a focus on the mathematical and statistical foundations of these fields.
- **AI Podcast**: A podcast that explores the latest developments in artificial intelligence and machine learning, with interviews and discussions with leading experts in the field.
- **Partially Derivative**: A podcast that covers data science and machine learning, with a focus on the practical applications of these technologies in business and industry.
- **Data Science at Home**: A podcast that covers machine learning and data science, with a focus on the tools and techniques that can be used to build and deploy machine learning models.
- **Data Science Radio**: A podcast that covers a wide range of data science and machine learning topics, with a focus on the practical applications of these technologies.

Links:

- Data Skeptic: https://dataskeptic.com/episodes
- Machine Learning Guide: https://www.machinelearningguide.net/
- Linear Digressions: https://www.lineardigressions.com/
- AI Podcast: https://lexfridman.com/ai/

- Partially Derivative: http://partiallyderivative.com/
- Data Science at Home: https://datascienceathome.com/
- Data Science Radio: https://datascienceradio.net/

RESEARCH PAPERS

Research Papers are an important resource for understanding the latest developments and advancements in the field of machine learning. They are written by experts in the field and present the results of their research and experiments in a formal and scholarly manner. Some popular websites to find research papers related to machine learning include arXiv, JMLR, and IEEE Xplore. Additionally, many universities and research institutions also have their own online repositories of research papers, which can be found by searching for the name of the institution followed by "research papers" or "publications." Some popular machine learning research papers include "A Few Useful Things to Know About Machine Learning" by Pedro Domingos, "Deep Residual Learning for Image Recognition" by Kaiming He, and "Generative Adversarial Networks" by Ian Goodfellow.

Links:

- arXiv: https://arxiv.org/
- IEEE Xplore Digital Library: https://ieeexplore.ieee.org/Xplore/home.jsp
- JSTOR: https://www.jstor.org/
- ResearchGate: https://www.researchgate.net/
- Google Scholar: https://scholar.google.com/

- Semantic Scholar: https://www.semanticscholar.org/

Reading and staying up to date with research papers is a great way to stay informed about the latest developments in the field of machine learning. These websites provide access to a wide range of papers from various sources, including journals, conferences, and preprint servers. Some of these websites also provide tools for searching and filtering papers based on specific keywords or authors. Additionally, many papers also provide links to the code and data used in the research, which can be useful for reproducing the results or building on the work.

Appendix C

C. TOOLS AND FRAMEWORKS

In this section, we will discuss various tools and frameworks that are commonly used in the field of machine learning. These tools and frameworks can help you to efficiently implement and test your machine learning models, and also help you to visualize and interpret your results.

Some popular tools and frameworks for machine learning include:

- **TensorFlow**: Developed by Google, TensorFlow is an open-source library for machine learning that allows users to build and train neural networks. It is widely used in deep learning, and is supported by a large community of developers.
- **Scikit-learn**: This is a popular machine learning library for Python that provides a wide range of tools for machine learning, including supervised and unsupervised learning algorithms, as well as pre-processing and model selection tools.
- **Keras**: This is an open-source library for deep learning that runs on top of TensorFlow. It provides a high-level, user-friendly interface for building neural networks.

- **PyTorch**: Developed by Facebook, PyTorch is an open-source machine learning library for Python that is particularly well-suited for deep learning.
- **Theano**: This is an open-source library for machine learning that allows users to define, optimize, and evaluate mathematical expressions involving multi-dimensional arrays.
- **R**: R is a programming language and software environment for statistical computing and graphics. It is widely used in the field of machine learning and is supported by a large community of developers.

These are just a few examples of the many tools and frameworks available for machine learning. Each has its own strengths and weaknesses, and the best one for your project will depend on your specific requirements.

Links:

- TensorFlow: https://www.tensorflow.org/
- Scikit-learn: https://scikit-learn.org/
- Keras: https://keras.io/
- PyTorch: https://pytorch.org/
- Theano: http://deeplearning.net/software/theano/
- R: https://www.r-project.org/

Appendix D

D. DATASETS

Datasets are an important resource for machine learning practitioners as they provide the necessary input data for training and evaluating models. There are many publicly available datasets that can be used for a wide range of tasks such as image classification, object detection, natural language processing, and time series forecasting. Some popular datasets include:

- MNIST: A dataset of handwritten digits for image classification tasks.
- CIFAR-10 and CIFAR-100: A dataset of natural images for image classification tasks.
- Imagenet: A dataset of over 14 million images for image classification and object detection tasks.
- UCI Machine Learning Repository: A collection of datasets for various tasks such as classification, regression, and clustering.
- Kaggle Datasets: A collection of datasets for various tasks, including many that are specific to certain industries or use cases, such as healthcare, finance, and natural language processing.
- Common Crawl: A dataset of over 25 billion web pages for natural language processing tasks.

- The Enron Email Corpus: A dataset of over half a million emails for text classification tasks.
- Time Series Data Library (TSDL): A dataset of over 600 univariate time series for time series forecasting tasks.

Links:

- MNIST: http://yann.lecun.com/exdb/mnist/
- CIFAR-10 and CIFAR-100: https://www.cs.toronto.edu/~kriz/cifar.html
- Imagenet: http://www.image-net.org/
- UCI Machine Learning Repository: http://archive.ics.uci.edu/ml/index.php
- Kaggle Datasets: https://www.kaggle.com/datasets
- Common Crawl: http://commoncrawl.org/
- The Enron Email Corpus: https://www.cs.cmu.edu/~enron/
- Time Series Data Library (TSDL): http://robjhyndman.com/TSDL/

OPEN-SOURCE DATASETS

- UCI Machine Learning Repository: https://archive.ics.uci.edu/ml/index.php
- Kaggle: https://www.kaggle.com/datasets
- OpenML: https://www.openml.org/search?type=data

- Quandl: https://www.quandl.com/search?query=machine+learning
- Data.gov: https://www.data.gov/topic/machine-learning
- Google's Dataset Search: https://datasetsearch.research.google.com/
- DataWorld: https://data.world/search?q=machine+learning
- Microsoft Research Open Data: https://msropendata.com/
- Amazon's Registry of Open Data on AWS: https://registry.opendata.aws/
- Data.gov.uk: https://data.gov.uk/search?q=machine+learning
- Data.gov.au: https://data.gov.au/search?q=machine+learning

Appendix E

E. CAREER RESOURCES

In the field of machine learning, there are a variety of resources available to help individuals advance their careers and stay up to date on the latest developments and trends. Some popular career resources include:

1. Kaggle: https://www.kaggle.com/ This website offers a wide variety of machine learning competitions and job listings for data scientists and machine learning engineers. It's a great way to gain experience, improve your skills and get noticed by potential employers.

2. Data Science Central: https://www.datasciencecentral.com/ This website is a great resource for data science and machine learning professionals, offering articles, tutorials, webinars, and job postings.

3. Data Science Society: https://www.datasciencesociety.net/ This community is a great resource for data science and machine learning professionals, providing a platform for networking, job searching and learning.

4. Data Science Collective: https://www.datasciencecollective.com/ This website is a great resource for data science and machine learning professionals, offering a wide variety of articles, tutorials and webinars on the latest trends and developments in the field.

5. Machine Learning Mastery: https://machinelearningmastery.com/ This website offers a wide variety of tutorials, courses and articles on machine learning, deep learning and artificial intelligence.

6. Data Science Master: https://datasciencemaster.com/ This website provides a range of resources to help individuals advance their careers in data science, machine learning and artificial intelligence, including job postings, tutorials and webinars.

7. Data Science Mastery: https://datasciencemastery.com/ This website offers a wide variety of tutorials, courses, and articles on data science and machine learning, to help individuals improve their skills and advance their careers.

Job Boards:

1. LinkedIn - https://www.linkedin.com/jobs/
2. Indeed - https://www.indeed.com/
3. Glassdoor - https://www.glassdoor.com/Job/jobs.htm
4. Kaggle - https://www.kaggle.com/jobs
5. SimplyHired - https://www.simplyhired.com/
6. Monster - https://www.monster.com/
7. CareerBuilder - https://www.careerbuilder.com/
8. The Muse - https://www.themuse.com/jobs
9. FlexJobs - https://www.flexjobs.com/

10. Dice - https://www.dice.com/

Note: These are some of the popular job boards for machine learning and data science roles, but it is always a good idea to also check out job listings on company websites and other relevant industry websites as well.

COMPANIES AND STARTUPS WORKING IN THE FIELD OF MACHINE LEARNING

1. Google
2. Amazon
3. Microsoft
4. Facebook
5. Apple
6. IBM
7. Intel
8. NVIDIA
9. OpenAI
10. DeepMind

These are just a few examples of large companies that have a significant presence in the field of machine learning. There are also many smaller startups and niche companies that specialize in specific areas of machine learning such as computer vision, natural language processing, or autonomous driving.

Links:

1. Google: https://www.google.com/about/careers/teams/research-science/
2. Amazon: https://www.amazon.jobs/en/teams/machine-learning

3. Microsoft: https://www.microsoft.com/en-us/research/research-area/artificial-intelligence/

4. Facebook: https://research.fb.com/category/machine-learning/

5. Apple: https://www.apple.com/jobs/us/teams/machine-learning.html

6. IBM: https://www.ibm.com/blogs/research/category/artificial-intelligence/

7. Intel: https://www.intel.com/content/www/us/en/artificial-intelligence/overview/artificial-intelligence-solutions.html

8. NVIDIA: https://www.nvidia.com/en-us/deep-learning-ai/industries/

9. OpenAI: https://openai.com/

10. DeepMind: https://deepmind.com/applied/

RESEARCH LABS AND UNIVERSITIES WITH A FOCUS ON MACHINE LEARNING

Research Labs and Universities with a focus on Machine Learning are organizations that conduct research and development in the field of machine learning. They often have teams of researchers and scholars working on various projects related to machine learning, such as developing new algorithms, analyzing large datasets, and creating new applications for machine learning. Some examples of research labs and universities with a focus on machine learning include:

- MIT Computer Science and Artificial Intelligence Laboratory (CSAIL)
- Stanford Artificial Intelligence Laboratory (SAIL)

- Carnegie Mellon University's Machine Learning Department
- The University of California, Berkeley's Artificial Intelligence Research Laboratory (BAIR)
- The University of Cambridge's Machine Learning Group
- The Max Planck Institute for Intelligent Systems
- The Alan Turing Institute for Data Science and Artificial Intelligence
- The Google Brain Team
- The Facebook AI Research (FAIR) Team
- The DeepMind Team

These organizations often collaborate with industry partners and publish their research in top-tier journals and conferences. They also frequently host workshops, seminars, and conferences to share their findings with the broader machine learning community.

GOVERNMENT ORGANIZATIONS AND FUNDING AGENCIES SUPPORTING ML RESEARCH AND DEVELOPMENT

- National Science Foundation (NSF)
- National Institutes of Health (NIH)
- Defense Advanced Research Projects Agency (DARPA)
- Intelligence Advanced Research Projects Activity (IARPA)
- National Aeronautics and Space Administration (NASA)

- National Oceanic and Atmospheric Administration (NOAA)
- Department of Energy (DOE)
- Department of Defense (DOD)
- Federal Aviation Administration (FAA)
- National Institute of Standards and Technology (NIST)
- National Institutes of Justice (NIJ)
- National Security Agency (NSA)
- National Geospatial-Intelligence Agency (NGA)
- Central Intelligence Agency (CIA)
- Federal Bureau of Investigation (FBI)
- Department of Homeland Security (DHS)
- Federal Emergency Management Agency (FEMA)
- United States Geological Survey (USGS)
- United States Army Research Laboratory (ARL)
- United States Navy Research Laboratory (NRL)
- United States Air Force Research Laboratory (AFRL)
- United States Marine Corps Warfighting Laboratory (MCWL)
- United States Coast Guard Research and Development Center (RDC)
- United States Special Operations Command (SOCOM)

- United States Cyber Command (CYBERCOM)
- United States Strategic Command (STRATCOM)
- United States Transportation Command (TRANSCOM)
- United States Space Command (SPACECOM)
- United States Joint Forces Command (JFCOM)
- United States Northern Command (NORTHCOM)
- United States Pacific Command (PACOM)
- United States Southern Command (SOUTHCOM)
- United States European Command (EUCOM)
- United States Africa Command (AFRICOM)
- United States Central Command (CENTCOM)
- United States Transportation Command (TRANSCOM)
- United States Strategic Command (STRATCOM)
- United States Cyber Command (CYBERCOM)
- United States Special Operations Command (SOCOM)
- United States Space Command (SPACECOM)
- United States Joint Forces Command (JFCOM)
- United States Northern Command (NORTHCOM)
- United States Pacific Command (PACOM)

Appendix F

F. GLOSSARY

A

- **Accuracy**: A measure of how well a model correctly predicts the outcomes of a dataset, typically measured as the ratio of correct predictions to total predictions.
- **Activation function**: A function applied to the output of a neuron in order to introduce non-linearity into the model.
- **Adaline**: Adaptive Linear Neuron, an algorithm developed by Bernard Widrow and Tedd Hoff in 1960, which is considered a precursor to modern artificial neural networks.
- **AUC-ROC**: Area Under the Receiver Operating Characteristic Curve, a metric used to evaluate the performance of binary classification models.
- **Autoencoder**: A type of neural network architecture in which the input and output layers have the same number of nodes, and the network is trained to reconstruct the input from the output.

B

- **Backpropagation**: An algorithm used to train artificial neural networks by iteratively adjusting the weights of the network in order to minimize the error between the predicted and actual outputs.
- **Batch Gradient Descent**: A variation of gradient descent algorithm where the parameters are updated after calculating the gradients of the loss function w.r.t to the parameters, from the entire training dataset.
- **Bias**: A term used to describe the difference between a model's predictions and the true values of the data.
- **Bias-Variance Trade-off**: A concept in machine learning that refers to the balancing act between a model's ability to fit the training data well (low bias) and its ability to generalize to new data (low variance).

C

- **Classification**: A type of supervised machine learning task in which the model is trained to predict a categorical label for a given input.
- **Clustering**: A type of unsupervised machine learning task in which the model is trained to group similar inputs together based on certain features.
- **Convolutional Neural Network (CNN):** A type of neural network architecture commonly used for image and video processing tasks.

D

- **Decision Tree**: A type of model used for both classification and regression tasks, in which the model is trained to make a series of decisions based on the input data.

- **Deep Learning**: A subfield of machine learning focused on building complex neural network architectures, such as convolutional neural networks and recurrent neural networks.

- **Density-Based Clustering**: A clustering algorithm in which clusters are defined as dense regions of the data space.

- **Dimensionality Reduction**: A technique used to reduce the number of features in a dataset by transforming the data into a lower-dimensional space.

E

- **Ensemble Methods**: A method of combining multiple models to improve the overall performance of the model.

- **Epoch**: A complete iteration over all the training data during the training of a model.

- **Extreme Gradient Boosting (XGBoost)**: An open-source library for gradient boosting, which is used to improve the performance of decision trees.

F

- **Feature**: A measurable property of the input data used to make predictions.

- **Feature Engineering**: The process of selecting and transforming features to improve the performance of a model.

- **Feature Scaling**: A technique used to normalize the range of features in a dataset.

- **F1 Score**: A metric used to evaluate the performance of classification models, which is the harmonic mean of precision and recall.

G

- **Gradient Descent**: An optimization algorithm used to minimize the loss function of a model by adjusting the model's parameters.
- **Gaussian Mixture Model (GMM)**: A generative probabilistic model that represents a set of data as a mixture of Gaussian distributions.

H

- **Hyperparameter**: A value that is set before training a machine learning model and is not learned during the training process. Examples include the learning rate and number of hidden layers in a neural network.

K

- **K-fold cross-validation**: A technique used to evaluate the performance of a model by dividing the data into k partitions, training on k-1 partitions, and evaluating on the remaining partition. This process is repeated k times, with each partition serving as the evaluation set once.

L

- **L1 regularization**: A technique used to prevent overfitting by adding a penalty term to the cost function that is proportional to the absolute value of the coefficients. This technique results in sparse solutions, with many coefficients equal to zero.

- **L2 regularization**: A technique used to prevent overfitting by adding a penalty term to the cost function that is proportional to the square of the coefficients. This technique results in small, non-zero coefficients.
- **Learning rate**: A hyperparameter that controls the step size in the optimization of a model's parameters. A small learning rate may result in slow convergence, while a large learning rate may overshoot the optimal solution.
- **Log loss**: A loss function used in classification tasks that measures the performance of a classifier by penalizing false classifications.
- **Label**: The output or target variable of a supervised learning problem.
- **Linear Regression**: A supervised learning algorithm used for predicting a continuous target variable based on one or more input features. Linear regression models the relationship between the input features and the target variable as a linear equation.
- **Logistic Regression**: A supervised learning algorithm used for classification problems, where the goal is to predict a binary outcome. Logistic regression models the probability of the positive class as a logistic function of the input features.
- **Loss Function**: A function used to evaluate the performance of a machine learning model, typically by penalizing the model for incorrect predictions. The goal of training a machine learning model is to minimize the loss function.

M

- **Machine Learning**: A subfield of artificial intelligence that involves the development of algorithms that can learn from data and make predictions or decisions without being explicitly programmed.
- **Model**: A representation of a system or problem that can be used to make predictions or decisions. In machine learning, models are trained on a dataset and are then used to make predictions on new, unseen data.
- **Metric**: A function used to evaluate the performance of a model, such as accuracy or F1 score.

O

- **Overfitting**: A common problem in machine learning, where a model is trained too well on the training data and performs poorly on new, unseen data. Overfitting occurs when a model is too complex and is able to memorize the training data instead of generalizing to new data.

P

- **Precision**: A metric used in classification problems to evaluate the number of true positive predictions made by a model, as a proportion of all positive predictions made.

R

- **Recall**: A metric used in classification problems to evaluate the number of true positive predictions made by a model, as a proportion of all actual positive instances in the dataset.
- **Regularization**: A technique used to prevent overfitting in machine learning models by adding a penalty term to the loss function that discourages large values of the model parameters.

- **Root Mean Squared Error (RMSE)**: A metric used to evaluate the performance of a regression model, calculated as the square root of the mean of the squared differences between the predicted and actual values.

S

- **Supervised Learning**: A type of machine learning where the goal is to learn a mapping from input features to output labels, using a labeled dataset.
- **Scikit-learn**: A popular Python library for machine learning that provides a consistent interface to a wide range of machine learning algorithms.

T

- **Test set**: A set of data used to evaluate the performance of a trained model, separate from the training and validation sets.
- **Training set**: A set of data used to train a model, separate from the test and validation sets.

U

- **Unsupervised Learning**: A type of machine learning where the goal is to discover hidden patterns or structure in an unlabeled dataset, without a specific target variable in mind.
- **Underfitting**: A situation in which a model is too simple to capture the underlying relationship in the data, resulting in poor performance on both the training and test sets.

V

- **Validation**: The process of evaluating a machine learning model on a separate dataset, after training, to estimate its performance on new, unseen data.
- **Variance**: A measure of the variability of a model's predictions for different training sets. A high variance model is sensitive to small fluctuations in the training data and is at risk of overfitting.

W

- **Weight**: A parameter learned by a model during training.

Z

- **Zero-one loss**: A loss function used in classification tasks that measures the performance of a classifier by counting the number of incorrect classifications.

Made in United States
North Haven, CT
06 November 2026

21769188R10276